Studies in Fuzziness and Soft Computing

Volume 344

Series editor

Janusz Kacprzyk, Polish Academy of Sciences, Warsaw, Poland
e-mail: kacprzyk@ibspan.waw.pl

About this Series

The series "Studies in Fuzziness and Soft Computing" contains publications on various topics in the area of soft computing, which include fuzzy sets, rough sets, neural networks, evolutionary computation, probabilistic and evidential reasoning, multi-valued logic, and related fields. The publications within "Studies in Fuzziness and Soft Computing" are primarily monographs and edited volumes. They cover significant recent developments in the field, both of a foundational and applicable character. An important feature of the series is its short publication time and world-wide distribution. This permits a rapid and broad dissemination of research results.

More information about this series at http://www.springer.com/series/2941

Janusz Kacprzyk · Dimitar Filev
Gleb Beliakov
Editors

Granular, Soft and Fuzzy Approaches for Intelligent Systems

Dedicated to Professor Ronald R. Yager

Editors
Janusz Kacprzyk
Systems Research Institute
Polish Academy of Sciences
Warsaw
Poland

Gleb Beliakov
Deakin University
Burwood, VIC
Australia

Dimitar Filev
Research and Innovation Center
Ford Motor Company
Dearborn, MI
USA

ISSN 1434-9922   ISSN 1860-0808   (electronic)
Studies in Fuzziness and Soft Computing
ISBN 978-3-319-82076-7   ISBN 978-3-319-40314-4   (eBook)
DOI 10.1007/978-3-319-40314-4

Printed on acid-free paper

This Springer imprint is published by Springer Nature
The registered company is Springer International Publishing AG Switzerland

To Professor Ronald R. Yager

*This book is a token of appreciation to
Prof. Ronald R. Yager for his great scientific
and scholarly achievements, long-time
service to the fuzzy logic and more generally
to the artificial and computational
intelligence communities. It is appreciation
for his original novel thinking and his*

groundbreaking ideas that have reshaped many already well-established fields of science and technology, and have initialized and triggered interest and research in many new ones. He has particularly motivated by providing tools to enable the computerized implementation of the various aspects of human cognition for decision-making and problem solving.

For many years he has been one of the remarkable personalities in the world of science, technology and applied mathematics. His long-time career has been characterized by the quest for scientific excellence and unquestionable integrity. This is what has characterized great minds, scholars and mentors for centuries, following a long academic tradition with roots that started in the first universities in the Middle Ages, and has been followed over the centuries by academics all over the world who have been aware that any compromise in quality, integrity, good practices and tradition in the academia can lead to detrimental effects, finally to self-destruction.

Professor Yager has pioneered new research directions in knowledge representation and processing, handling of all kinds of imperfect information, imprecision, uncertainty and incompleteness. He has ingeniously used tools and techniques from both well-established areas, like the traditional probability theory and statistics, and many new emerging ones like fuzzy logic, possibility theory, the Dempster–Shafer theory and the theory of approximate reasoning. He has started new research areas that have, since his seminal papers, developed rapidly and have become objects of interest for hundreds or even thousands of scientists and scholars from all over the world. One can here mention, just to quote a few, his famous works on the OWA (ordered weighted averaging) operators, fuzzy quantifier driven aggregation of pieces of evidence, linguistic data summarization and participatory learning. He has provided a deeper understanding of fuzzy systems modeling, and developed a framework for using granular computing for modeling social networks. He has done considerable work in multi-criteria decision-making and decision analysis in the face of uncertainty. One must always be aware that even if his models have been just conceptual or theoretical at the first glance, this has been a judgment simply based only on appearances as they all have been profoundly based on realistic assumptions, and—as a result of those—have been applicable to a huge variety of relevant real-world technological, economic, financial, social and diverse problems.

If one looks carefully at what contributions Prof. Yager has made and when he has made these contributions, one can clearly see him as a visionary who has been able to earlier than most understand what is going to play an important role in science in the future, and then being able to put the pieces of ideas, tools and techniques together in an innovative way that has initiated new things and added value.

The world scientific and technology community has fully appreciated his great achievements and he has since the very *beginning of his research and scholarly activities been awarded with the highest honors, awards and prizes from the most influential and opinion making institutions and organizations from all over the world. They are too many to mention but this has culminated with the 2016 IEEE Frank Rosenblatt Award, one of the most prestigious distinctions for scientists and scholars working in the field of computational intelligence. The Rosenblatt Award Committee has fully appreciated his groundbreaking innovative ideas, but also his role as a leader for the research community, one who has been able to see and communicate new vistas, challenges and opportunities.*

The scientific and scholarly virtues of Prof. Yager mentioned above, as great and important as they can be, are not rich enough to fully characterize him. Behind every human being—no matter who, and how prominent or influential he or she is—is just a human being with his or her personality. Professor Yager is in this respect is a remarkable person who has been known for his consideration, generosity and respect, and modesty that is characteristic of great people. This all has always implied that there has been a good atmosphere around him, and has for sure contributed to his exceptional ability to inspire people around him.

Some of the editors of this volume have been privileged because they have had an opportunity to stay with Prof. Yager for a longer time, in different periods, starting from the early 1980s, through the 1990s to the 2000s. They have fully appreciated how inspiring contacts with him have been, and how their professional development has been shaped by discussions and daily interactions with him. They have also been able to feel his friendliness, integrity and great personal qualities.

It is clear that this volume, meant to be just a token of appreciation for Prof. Yager by our entire community, is small in comparison with what he has done in science, education, and what he has done to virtually all of us by both invaluable inspiration, but also friendship. We are honored to have had the opportunity to prepare this volume.

Janusz Kacprzyk
Dimitar Filev
Gleb Beliakov

Preface

This volume is a result of a special project the purpose of which was twofold. First of all, from a substantial point of view, we wished to provide a bird's view of some novel directions in the broadly perceived "intelligent systems", starting with more philosophical and foundational considerations, through a bunch of promising models for the analysis of data, decision-making, systems modeling, and control, with a message that an approach on the broadly perceived granular computing, soft computing and fuzzy logic can provide in this context some breakthrough views, perspectives, and—as a consequence—solutions, not only relevant from a conceptual and analytic points of view, but also practical, with a high implementation potential.

This book is basically a token of appreciation to Prof. Ronald R. Yager for his great scientific and scholarly achievements, long-time service to the fuzzy logic and more generally to the artificial and computational intelligence communities. It is appreciation for his original novel thinking and his groundbreaking ideas that have reshaped many already well-established fields of science and technology, and have initialized and triggered interest and research in many new ones. He has particularly motivated by providing tools to enable the computerized implementation of the various aspects of human cognition for decision-making and problem solving.

The second purpose of this volume has been to acknowledge the role and contributions of Prof. Ronald R. Yager in all these areas of modern science and technology. In his illustrious research and scholarly career, that has spanned over some decades, he has greatly contributed to the clarification of many foundational issues, proposed new methodological and algorithmic solutions, and has also been instrumental in their real-world applications. His original results cover both contributions in traditional well-established areas like probability theory and statistics, and many new emerging ones like fuzzy logic, possibility theory, the Dempster-Shafer theory, the theory of approximate reasoning, computing with words, to name a few. He has triggered research in new areas related to aggregation and fusion of information, linguistic data summarization, participatory learning, granular computing for systems modeling, etc. More information will be provided in the dedication part of this volume.

The editors of this volume have a special relation to Prof. Yager because they have spent many years with him as visiting professors and close collaborators. They have been very much privileged to be able to work in the early years of their careers with him, and enjoy both inspiration, wise advice, and also his friendship. This volume is a token of appreciation for him. In fact, practically all authors of contributions included in this volume have also had a very close and fruitful professional relation with Prof. Yager. He has also inspired them, and the dedication of their papers to him is also a proof of their deep appreciation for his great results and role in shaping their scientific interests and careers.

Of course, this volume, meant to be just a token of appreciation for Prof. Yager by our entire community, is small in comparison with what he has done in science, education, and what he has done to virtually all of us by both invaluable inspiration and friendship.

An important part of this volume is "Dedication" in which the contributions, research record, influence, etc. of Prof. Yager have been presented in more detail.

The volume starts with "Part I: Information theoretic and aggregation issues" which contains more basic and foundational contributions of the authors to this volume. They cover some most crucial problems and issues in the scope of this volume.

Enric Trillas and Rudolf Seising ("On the meaning and the measuring of 'probable'") are concerned with some basic issues related to probability, from the point of view of probability theory which is a mathematical theory that is an important part of pure mathematics, and as a long and distinguished history of more than 300 years. It has found a multitude of applications in almost all domains of science and technology. Then, the authors relate this to a relatively short history of fuzzy sets theory of just the lasts 50 years during which it has been theoretically developed and successfully applied in many fields. Then the authors, being aware of some controversies on the nature of fuzzy sets viewed in relation to probability. The paper's purpose is to provide a contribution to the clarification of some differences between fuzzy sets and probabilities, as viewed by the authors.

Didier Dubois, Henri Prade and Agnès Ricó ("Organizing families of aggregation operators into a cube of opposition") are concerned with the so-called cube of opposition which is a structure that extends the traditional square of opposition, known since the ancient times and widely employed in the study of syllogisms. The cube of opposition, which has recently been generalized to non-Boolean, graded statements, is shown in this paper to be applicable to well-known families of idempotent, monotonically increasing aggregation operations, for instance, used in multi-criteria decision-making, which qualitatively or quantitatively provide evaluations between the minimum and the maximum of the aggregated quantities. Some notable examples are here the weighted minimum and maximum, and more generally the Sugeno integrals on the qualitative side, and the Choquet integrals, with the important particular case of the OWA operators, on the quantitative side.

The main advantage of the cube of opposition is its capability to display various possible aggregation attitudes and to show their complementarity.

Bernadette Bouchon-Meunier and Christophe Marsala ("Entropy measures and views of information") consider various issues related to entropies and other information measures, relating to some extent their analysis to what Prof. Yager has done. They take into account the very concept of a particular type of a set in question in order to point out a similarity between the quantities introduced in various frameworks to evaluate a kind of entropy. They define the concept of an entropy measure and we show its main characteristics, mainly in the form of monotonicity which are satisfied by the ideas pioneered in this context by Yager.

H. Bustince, J. Fernandez, L. De Miguel, E. Barranechea, M. Pagola, and R. Mesiar ("OWA operators and Choquet integrals in the interval-valued setting") use the notion of an admissible order between intervals to extend the definition of the OWA operators and the Choquet integrals to the interval-valued setting. Then, using this more general and comprehensive setting, the authors present an algorithm for decision-making based on their new concepts and algorithms.

Paul Elmore and Frederick Petry ("Information Theory Applications in Soft Computing") provide an overview of information theoretic metrics and the ranges of their values for extreme probability cases. They heavily relate their analysis to imprecise database models including similarity-based fuzzy models and rough set models. More specifically, they show various entropy measures for these database models' content and responses to querying. Moreover, they discuss the aggregation of uncertainty representations, in particular the possibilistic conditioning of probability aggregation by using information measures to compare the resultant conditioned probability to the original probability for three cases of possibility distributions.

The second part of the volume, Part II: "Applications in modeling, decision making, control, and other areas", provides an account of various applications of modern tools and techniques of broadly perceived intelligent systems, computer science, decision analysis, etc., to formulate and solve many important practical problems.

Uzay Kaymak ("On practical applicability of the generalized averaging operator in fuzzy decision making") provides a deep and constructive analysis of, first, general issues related to the use of aggregation operators in decision-making, and then—more specifically—to the us of the generalized averaging operator as decision functions in the modeling human decision behavior in the context of decision-making. He uses real data to analyze the models discussed and provides a comparison with the results obtained by using compensatory operators. The numerical data suggests that the generalized averaging operator is well suited for the modeling of human decision behavior.

Leandro Maciel, Rosangela Ballini and Fernando Gomide ("Evolving possibilistic fuzzy modeling and application in value-at-risk estimation") propose an evolving possibilistic fuzzy modeling approach for value-at-risk modeling and estimation. Their approach is based on an extension of the possibilistic fuzzy c-means clustering and functional fuzzy rule-based systems. It employs memberships and typicalities to update clusters centers and forms new clusters using a statistical control distance-based criteria. The paradigm of evolving possibilistic fuzzy modeling (ePFM) also makes use of a utility measure to evaluate the quality of the current cluster structure which implies the fuzzy rule-based model. The authors are concerned with the market risk exposure which plays a key role for financial institutions in risk assessment and management, and use as a means to measure the risk exposure by evaluating the losses likely to incur when the prices of the portfolio assets decline. The value-at-risk (VaR) estimate is one of the most widely used measures of financial downside market risk, and the authors in the computational experiments evaluate the ePFM for the value-at-risk estimation using data of the main equity market indexes of United States (S&P 500) and Brazil (Ibovespa) from January 2000 to December 2012, and the econometric models benchmarks such as GARCH and EWMA, and state-of-the-art evolving approaches, are also compared against the ePFM. The results suggest that the ePFM is a potentially good candidate for the VaR modeling and estimation.

Janusz Kacprzyk, Hannu Nurmi and Sławomir Zadrożny ("Using similarity and dissimilarity measures of binary patterns for the comparison of voting procedures") consider an interesting and important problem of how similar and/or dissimilar voting procedures (social choice functions) are. They first extend their rough set based qualitative-type approach which makes it possible to partition the set of voting procedures considered into some subsets within which the voting procedures are indistinguishable, i.e., (very) similar. Then, they propose an extension towards a quantitative evaluation via the use of degrees of similarity and dissimilarity, not necessarily metrics and dual. The authors consider the following voting procedures: amendment, Copeland, Dodgson, max-min, plurality, Borda, approval, runoff and Nanson, and the following criteria Condorcet winner, Condorcet loser, majority winner, monotonicity, weak Pareto winner, consistency, and heritage. The satisfaction or dissatisfaction of the particular criteria by the particular voting procedures is represented as binary vectors. The similarity and dissimilarity measures of: Jaccard–Needham, Dice, Correlation, Yule, Russell–Rao, Sokal–Michener, Rodgers–Tanimoto, and Kulczyński are employed. The approach is shown to yield much insight into the similarity/dissimilarity of voting procedures.

Gloria Bordogna, Simone Sterlacchini, Paolo Arcaini, Giacomo Cappellini, Mattia Cugini, Elisabetta Mangioni, and Chrysanthi Polyzoni ("A geo-spatial data infrastructure for flexible discovery, retrieval and fusion of scenario maps in preparedness of emergency") are concerned with the following problem. In order to effectively plan both preparedness and response to emergency situations it is necessary to access and analyze timely information on plausible scenarios of occurrence of dangerous events. They use the so-called scenario maps which represent the estimated susceptibility, hazard or risk of occurrence of an event on a territory. Their generalization to real time is unfortunately difficult. Moreover, the application of physical or statistical models using environmental parameters representing current dynamic conditions is time-consuming and numerically demanding. To overcome these difficulties the authors propose an offline generation of scenario maps under diversified environmental dynamic parameters, and a geo-Spatial Data

Infrastructure (SDI) to allow people in charge of emergency preparedness and response activities to flexibly discover, retrieve, fuse and visualize the most plausible scenarios that may happen given some ongoing or forecasted dynamic conditions influencing the event. The solution proposed is novel in that it provides an ability to interpret flexible queries in order to retrieve risk scenario maps that are related to the current situation and to show the most plausible worst and best scenarios that may occur in each elementary area of the territory. A prototypical implementation concerns the use of scenarios maps for wild fire events.

Dimitar Filev and Hao Ying ("The multiple facets of fuzzy controllers: look-up-tables—A special class of fuzzy controllers") deal with look-up table (LUT) controllers which are among the most widely known and employed control tools in practice due to their conceptual simplicity, ease of use, inexpensive hardware implementation. Moreover, strong nonlinearity and multimodal behaviors can easily be handled by such controllers in many cases, only by experimentally measured data. The authors, in their previous paper, showed that the two-dimensional (2D) LUT controllers and one special type of two-input Mamdani fuzzy controllers are related as they have the identical input–output mathematical relation, demonstrated how to represent the LUT controllers by the fuzzy controllers, and showed how to determine the local stability of the LUT control systems. In the present work, they extend these results to the n-dimensional LUT controllers and the special type of the n-input Mamdani fuzzy controllers.

Pablo J. Villacorta, Carlos A. Rabelo, David A. Pelta and José Luis Verdegay ("FuzzyLP: an R package for solving fuzzy linear programming problems") consider fuzzy linear programming which is meant to overcome some inherent limitation of the traditional, widely used linear programming in which we need to know precisely all the conditions and parameters of the problem modeled. Since this is not always possible, a suitable alternate solution can be fuzzy linear programming which makes it possible to use imprecise data and constraints. The authors try to overcome a serious deficiency in that, in spite of a three decade long existence of fuzzy linear programming, there is still a serious difficulty in its proliferation, namely a lack of software developed for free use. The authors present an open-source R package to deal with fuzzy constraints, fuzzy costs and fuzzy coefficients in linear programming. First, the theoretical foundations for solving each type of problem are introduced, and then examples of the code. The package is accompanied by a user manual and can be freely downloaded, used and modified by any R user.

The last part of the volume, Part III: "Some bibliometric remarks", is somewhat unorthodox and special. It includes one paper which presents a detailed bibliometric analysis of main contributions of Prof. Yager, and their influence on research activities of many people from various areas who have been prolific and have developed their field of interest, to a large extent thanks to Yager's inspiration. This unusual part of the volume is fully justified by an extraordinarily high publication record and its wide recognition of Yager.

More specifically, José M. Merigó, Anna M. Gil-Lafuente and Janusz Kacprzyk ("A bibliometric analysis of the publications of Ronald R. Yager") present a bibliometric analysis of the vast publications record of Prof. Ronald R Yager. They use the data available in the Web of Science where he has more than 500 publications. This bibliometric review considers a wide range of issues including a specific analysis of his publications, collaborators and citations. A novel use of a viewer software is used to visualize his publication and citation network though bibliographic coupling and a co-citation analysis. The results clearly show his strong influence in computer science, although it also shows his strong influence in engineering and applied mathematics too.

We wish to thank all the contributors to this volume. We hope that their papers, which consititute a synergistic combination of foundational and application-oriented works, including relevant real-world implementations, will be interesting and useful for a large audience.

We also wish to thank Dr. Tom Ditzinger, Dr. Leontina di Cecco, and Mr. Holger Schaepe from Springer for their dedication and help to implement and finish this publication project on time maintaining the highest publication standards.

Warsaw, Poland   Janusz Kacprzyk
Dearborn, USA   Dimitar Filev
Burwood, Australia   Gleb Beliakov
April 2016

Contents

Part I
Information Theoretic
and Aggregation Issues

On the Meaning and the Measuring of 'Probable'

Enric Trillas and Rudolf Seising

To Professor Ron.R. Yager, with deep affection.

Abstract Probability has, as a mathematical theory that is an important part of pure mathematics, a long and distinguished history of more than 300 years, with fertile applications in almost all domains of science and technology; but the history of fuzzy sets only lasts 50 years, during which it was theoretically developed and successfully applied in many fields. From the very beginning there was, and there still is, a controversy on the nature of fuzzy sets viewed by its researchers far from randomness, and instead close by probabilists. This paper only goal is nothing else than trying to contribute to the clarification on the differences its authors see between fuzzy sets and probabilities and through the representation, or scientific domestication, of meaning by quantities.

Keywords Meaning's representation · Probable · Imprecision · Uncertainty · Fuzzy probability

1 Introduction

The predicate P = probable, used in both Natural Language and Science, scarcely deserved a careful and specific study of its linguistic meaning and its measuring either by numbers, or by fuzzy sets. It is also scarcely studied the use of the words

E. Trillas
European Centre for Soft Computing, Mieres (Asturias), Spain

R. Seising (✉)
Institute of History of Medicine, Natural Sciences and Technology
"Ernst-Haeckel-Haus", Friedrich Schiller University Jena, Jena, Germany
e-mail: rudolf.markus.seising@uni-jena.de

© Springer International Publishing Switzerland 2017
J. Kacprzyk et al. (eds.), *Granular, Soft and Fuzzy Approaches for Intelligent Systems*, Studies in Fuzziness and Soft Computing 344,
DOI 10.1007/978-3-319-40314-4_1

not-probable and improbable, without whose use it is difficult, if not impossible, to learn how the word probable [20] can be used, and that are sometimes confused (also in dictionaries); how can it be recognized that something is probable without recognizing that some other thing is improbable? People usually learn how to use words by polarity, that is, by jointly learning the use of a word and that of either its negation, or its antonym.

In the next section of this introduction we give some remarks on the history and philosophy of the concept of probability and the calculus of probabilities. Then, after the typically scientific 'domestication of meaning' by quantities, explained in some papers by the first author [21], it seems possible to begin with a systematic study of such goal. This is what this paper just tries to offer by considering the linguistic predicate P = probable in a generic universe of discourse X, not previously endowed with an algebraic structure, and provided it can be recognized the two arguments' relationship 'less probable than', naturally associated to all imprecise predicate, or linguistic label. Recognizing this relationship is a right way for asserting that 'probable' is a gradable predicate.

Let us designate such empirically perceived relationship by the binary relation

$$\leq_P \subseteq X \times X : x \leq_P y \Leftrightarrow x \text{ is less probable than } y.$$

With this relation the previously amorphous universe X is endowed with the graph $(X, \leq_P)$ structure, representing the qualitative meaning of P in X. Without knowing this graph it is not possible to study up to which extent the elements in X are actually probable, that is, the full meaning of P in X. It should be noticed that such relations $\leq_P$ only reduce to $=_P$ (equally 'P' than) provided the predicate 'P' is precise, or rigid, in X [21].

1.1

When Pierre-Simon de Laplace (1749–1827) first published his *Essai philosophique sur les probabilités* in the year 1814, he gave distinction to the philosophical world view of strict physical determinism:

> We may regard the present state of the universe as the effect of its past and the cause of its future. An intellect which at a certain moment would know all forces that set nature in motion, and all positions of all items of which nature is composed, if this intellect were also vast enough to submit these data to analysis, it would embrace in a single formula the movements of the greatest bodies of the universe and those of the tiniest atom; for such an intellect nothing would be uncertain and the future just like the past would be present before its eyes. [12, p. 4]

Laplace was inspired by the classical theory of mechanics, created by Isaac Newton (1642–1727), that allows us to predict the motions of all things based on knowledge of their starting points, their velocities, and the forces between them. He also knew the *Principle of Sufficient Reason* that Gottfried Wilhelm Leibniz (1646–1716) had already postulated in 1695:

> Everything proceeds mathematically … if someone could have a sufficient insight into the
> inner parts of things, and in addition had remembrance and intelligence enough to consider
> all the circumstances and take them into account, he would be a prophet and see the future
> in the present as in a mirror. [13]

Already the ancient Greek philosopher Aristotle (384 BC 322 BC) had written:
"All men by nature desire knowledge" [2]. How do we do this?—First of all we
use our perceptions, he said, second we use retrospection, and third we use our
experiences. What we call arts, technique and science appears on the basis of our
experiences. However, scientific knowledge will never be perfect, as Laplace put
on record when he added the following sentence to the already quoted paragraph
to make clear that his assumed all-knowing intellect—later it was named *Laplace's
demon*—cannot be a human:

> The human mind offers, in the perfection which it has been able to give to astronomy, a
> feeble idea of this intelligence. [12, p. 4]

Even all our scientific efforts, "in the search for truth", as Laplace wrote, "tend to
lead it back continually to the vast intelligence which we have just mentioned, but
from which it will always remain infinitely removed" [12, p. 4].

Laplace was convinced that we need the concept of probability to bridge this
gap. Because we will never know "all forces that set nature in motion, and all positions of all items of which nature is composed" and therefore "the future just like
the past of all items of which nature is composed" will "never be present before our
human eyes"! In his already in 1878 published *Memoire sur les probabilités*, he illus-
trated this as follows: "Each event being determined by virtue of the general laws of
this universe, it is probable only relatively to us, and, for this reason, the distinction of its absolute possibility and of its relative possibility can seem imaginary."
[11, p. 2] His thinking on a philosophical concept and a calculus of probabilities was one milestone to this branch of mathematics. However, its history is
much older: we can trace it back at least until the works of Gerolamo Cardano
(1501–1576), Christiaan Huygens (1629–1695), Blaise Pascal (1623–1662) and
Pierre de Fermat (1601/7–1665). Today, we consider "Probability theory" to be an
axiomatized mathematical theory as it is in the 1933 published axiom system of
Andrey N. Kolmogorov (1903–1987) that provides a calculus to treat uncertainties
about the outcomes of random phenomena [9].

1.2

There are two basic philosophical views on classical probability, namely the frequen-
tist and the Bayesian [6]. The first, views probability as a limit long-run frequency of
outcomes occurrences, and the second, as a degree of belief in the occurrence of an
outcome. Both show advantages and disadvantages, but, anyway, the following two
linguistic components of fuzzy sets must be carefully considered in order to establish
a probability for them: the semantic and the syntactic components. The first corresponds to the information on the objects necessary for specifying 'events' that can
be predicated as probable; it concerns to establish a suitable definition extending the
classical concept of measurable-space, or sigma-algebra, to a fuzzy set algebra [1].

The second corresponds to the algebraic places in which probable events, or statements, can be represented and probabilities assigned to such representations; for it, the essential properties of such probabilities, that should allow computations with them, must be established. For what concerns the 'abstract' definition of a probability, this paper basically only takes into account that of Kolmogorov, and as it is presented by Kappos [8] in the setting of Boolean algebras, that has generated the mathematical theory of probability in the form in which it is currently considered nowadays, and to which the great successes reached in the last 80 years are in debt. In general, and to scientifically domesticate any predicate, that is, representing it by means of quantities [25], both semantic and syntactic aspects should be considered; there is no scientific theory without a formal frame of representation, and theories are empty if they do not refer to something meaningful in a well established ground.

2   The Measure of P = Probable

2.1

Let $(L, \leq)$ be a poset with extreme elements 0 (minimum), and 1 (maximum). An L-measure for the graph $(X, \leq_P)$ is a mapping $p : X \to L$, such that:

(a) If $x \leq_P y$, then $p(x) \leq p(y)$
(b) If z is a minimal for $\leq_P$, then $p(z) = 0$
(c) If z is maximal for $\leq_P$, then $p(z) = 1$.

Once an L-measure p is known, it is said that each triplet $(X, \leq_P, p)$ reflects a meaning of P in X, and L should be chosen accordingly with the purposes and limitations of the subject into consideration. For what concerns the predicate 'probable' it is important to consider the cases in which L is either the unit interval $[0, 1]$ of numbers in the real line (classical case), or the set $[0, 1]^{[0,1]}$ of fuzzy sets, in the same interval (fuzzy case). The first is totally ordered by the usual linear order of the real line, and the second is only partially ordered by the pointwise order between fuzzy sets, extending the first by,

$$\mu \leq \sigma \Leftrightarrow \mu(x) \leq \sigma(x), \text{ for all } x \text{ in } [0, 1];$$

the extremes of the first are the numbers 0 and 1, and those of the second are the fuzzy sets $\mathbf{0}(x) = 0$, and $\mathbf{1}(x) = 1$, for all x in $[0, 1]$; both are posets. In principle, the universe of discourse X is any set to whose elements it is applicable the predicate 'probable' that from now on will be designated by P, and its measures by p; it should be noticed that, in general, neither $(X, \leq_P)$, nor $(L, \leq)$, are lattices.

The measures p of P = probable, are called probabilities, and if X is a set of fuzzy sets, it is said that p is a 'fuzzy-crisp' or a 'fuzzy-fuzzy' probability, provided that, respectively, L is the unit interval, or the set $[0, 1]^{[0,1]}$ of 'fuzzy numbers'.

2.2

The classical mathematical theory of probability is developed under the Kolmogorov idea, here shortened as the crisp-crisp probability, and on the hypotheses that the universe of discourse X is endowed with either the algebraic structure of an Orthomodular lattice (Kolmogorov's Quantum Probability [5]), or with the particular structure of a Boolean algebra (strict Kolomogorov's Probability [10]), but always supposing that the probability ranges in the unit interval, or in some special cases, in the complex field. Let us stop in these cases for a while, and designing the algebraic structure by $\Omega = (X; 0, 1; \cdot, +,')$, whose natural partial order corresponds to its lattice [4] character: $a \leq b \Leftrightarrow a \cdot b = a \Leftrightarrow a + b = b$. With this 'natural order', 0 is the minimum and 1 the maximum of X in Ω.

In both cases, it is always supposed that the relation 'a is less probable than b' coincides with $a \leq b$ [24], that is, $\leq_P = \leq$, and, that provided, in the Boolean algebra case, $\leq$ is nothing else than the set's contention $\subseteq$. The idea behind this hypothesis is that a set A is less probable than a set B if and only if the second is bigger or equal to the first (has more elements), something that comes from the empirical fact that, the more black balls an urn with a fixed number of other color balls has, the more probable is to extract a black ball from it. This hypothesis corresponds to making 'probable' dependent from the cardinality of the set of elements to which it is applied, and it is behind the first known definition of probability: prob $(A) =$ (number of elements in A/(total number of elements in X) or, as it is usually shortened, (favorable cases)/(total cases). Obviously, if neither the elements in Ω are crisp sets, nor the problem is a kind of 'counting' type, it is difficult to adopt such hypothesis. For instance, and as it is shown by the *Sorites* methodology [23], the linguistic predicate 'tall' cannot be represented by a crisp set, and in a statement like 'John is tall' no way of counting seems to exist 'a priori'; hence there is no a frequentistic way for approaching the crisp-probability that John is tall, and the only that seems actually possible is to describe the predicate 'tall' by its membership function in a given universe of discourse, by making it dependent, for instance, on the centimeters of tall of its inhabitants. If the shortest one measures 130 cm, the taller is 195 cm, and John is 170 cm, it does not seem bizarre to comparatively say that John is tall up to the degree 0.6.

2.3

A probability p in Ω is not directly defined as a measure, as they are defined in "2.1", but is defined by the Kolmogorov's basic axioms [8, 10],

1. $p(X) = 1$,
2. If $a \leq b'$ (b is contradictory with a), then $p(a + b) = p(a) + p(b)$, from which it follows,

   - $p(a') = 1 - p(a)$, since a' is contradictory with $a(a = a'' = (a')')$, and hence $p(a + a') = p(X) = 1 = p(a) + p(a')$. Thus $(0) = p(X') = 1 - 1 = 0$,
   - If Ω is a Boolean algebra, since it is $a \cdot b = 0 \Leftrightarrow a \leq b'$, (2) can be re-stated by,
   - 2^*. If $a \cdot b = 0$, then $p(a + b) = p(a) + p(b)$.

Then, and provided it is a $\leq b \Leftrightarrow b = a + b \cdot a'$ (orthomodular law [4]), and $a \leq (b \cdot a')' = b' + a$, from $p(b \cdot a') \geq 0$, it follows: $p(b) = p(a) + p(b \cdot a') \geq p(a)$. Thus, p is a measure for P in Ω in the sense of "2.1". Of course, this does not mean that all those measures are probabilities.

Remarks

   I. It should be pointed out that, in Ortholattices, $a \leq b'$ implies $a \cdot b \leq b' \cdot b' = 0$, that is, $a \cdot b = 0$: contradiction implies incompatibility. The reciprocal is only true in the particular case of Boolean algebras; only in this case, and for all a and b, von Neumann's law [4] of 'perfect repartition' holds: $a = a \cdot b + a \cdot b'$ [∗]. Than, if $a \cdot b = 0$, it follows that $a = a \cdot b' \Leftrightarrow a \leq b'$ and, hence, contradiction and incompatibility are equivalent. In Boolean algebras there is no way of distinguishing both concepts. In Orthomodular lattices both concepts are equivalent only for pairs of orthogonal elements a and b, that is, those that verify the former equality [∗].

  II. In the case of De Morgan algebras, of which Boolean algebras are also a particular case, and in general, there is no any implication between incompatibility and contradiction, although there are cases in which one of them exists., as it can be shown in [0, 1] if endowed with a t-norm, a t-conorm, and a strong negation [26]. In principle, they remain independent properties. Analogously, with fuzzy sets and replaying what happens in language, contradiction and incompatibility are independent properties. Hence, if reproducing the axioms 1., 2., or 1., 2*. with fuzzy sets, proving that a probability is a measure can be done only in some particular cases and with those algebras of fuzzy sets allowing it.

 III. Provided X is neither an Orthomodular lattice, nor a Boolean algebra, but, as it is the case in language, a previously unstructured set of linguistic statements, then, for instance, to state that 'x is P is less probable than y is Q', requires to specify this binary relationship. Provided the statements can be represented by fuzzy sets, it should be established the corresponding binary relation between them that, were it not the typical pointwise order, makes impossible, or at least very difficult, a formalization in algebras of fuzzy sets like the standard ones. For this reason, and analogously to the classic case, it is also often supposed that $\leq_P$ equals or, at least, is contained in the pointwise order of fuzzy sets; anyway, in all cases this hypothesis should be checked with the actual problem's conditions. Notice that, in the sets' case, $A \subseteq B$ actually seems to imply that A is less probable than B but, provided the sets were not finite, the reciprocal does not seem always immediately clear and then the former hypothesis can be doubtful.

  IV. Any measure of probable requires a suitable definition of 'conditional probability' [19] to compute which is the probability of an event which is conditioned by another one or, plainly, the first is unthinkable without the previous occurrence of the second. That is, the case in which the measure of an event a only has sense under another event b, through the conditional statement 'if b, then a'. In the Kolmogorov theory, conditional probability is defined by

$$p(a/b) = p(a \cdot b)/p(b) \text{ provided } p(b) > 0,$$

that comes from the supposition that the conditional statement 'if b, then a' just means 'a and b', that is, it is represented in the conjunctive form '$a \cdot b$'. Since it is $p(a/b) = p(a \cdot b/b)$, the one-argument function $p(\cdot/b)$ is actually a measure of probability but only in the new algebra X/b, whose elements are $x \cdot b$ ($x \cap B$, with sets), for all x in the given Boolean algebra X, with minimum and maximum $0 = 0 \cdot b$ (or $\emptyset = \emptyset \cap A$), and $b = 1 \cdot b$ ($B = X \cap B$), and whose complement is $a^* = b \cdot a'$ ($A^* = B \cap A^c$). This new algebra can be called the algebra 'diaphragmed' by b, and whose maximum and minimum are, respectively, b and 0, where it is $p(b/b) = 1$, $p(0/b) = 0$, $a \leq c \Rightarrow p(a/b) \leq p(c/b)$, $p(a^*/b) = 1 - p(a/b)$, etc. Notice that X/b coincides with X if and only if it is $a' = a' \cdot b \Leftrightarrow a' \leq b$, for all a in X, that implies $b = 1$; there is only coincidence if $b = 1$. From the linguistic point of view, to use the conditional probability as defined before it is necessary that conditional statements accept a representation in the 'conjunctive' form. From conditional probability it can said that an event a is p-independent of another event b, provided $p(a/b) = p(a) \Leftrightarrow p(a \cdot b) = p(a)$.

V. Objects (physical or not) to which the predicate 'probable' is applied and a measure of it can be found, are traditionally called 'events', a name coming from the 'eventuality' of something, for instance, of obtaining 'four' in throwing a dice. Events (a word whose etimology lies in the Latin 'eventus': something coming from outside) are those elements in an Orthomodular lattice, sets in a Boolean algebra, fuzzy sets in a set, or statements in a universe of discourse, to which the predicate $P = $ probable can be applied and measured; that is, for which quantities $(X, \leq_P, p)$ can be known. Probable should not be confused with provable, that refers to conjecturing that something will admit a proof.
Something is eventual, or contingent, if it can be thought but in whose occurrence or reality there is no a total safety: Eventually, four could be obtained; eventually, the variable could take values between 5 and 10; eventually, a football player could be short, etc. What is eventual is but a conjecture that can be probable, even if it is not known how to proceed to measure up to which extent it is so. Also the concept of plausible is different, referring to the worthiness of happening: It is plausible that such event is probable, it is plausible that such mathematical conjecture could be proven false, etc.

VI. The three axioms defining a measure p, are insufficient to specify a single p; more information is needed for obtaining each one of them. For instance, in the experiment of throwing a dice, to know the measure, or probability, of an event, for instance 'getting five', it is necessary either to know some characteristics of the experiment, for instance, if the dice is a perfect cube, if the dice is internally charged, if the launching surface is plane and not elastic, or it is with sand, etc., or, alternatively, to suppose the hypothesis that it is an 'ideal' dice. The two axioms of a probability just conduct to the equality $1 = p_1 + p_2 + \cdots + p_6$, with $p_i = $ measure of the event 'getting i', for i from 1 to 6, belonging to $[0, 1]$, and with which it is $p_5 = 1 - (p_1 + p_2 + p_3 + p_4 + p_6)$ making necessary to know the values of the p_k in the parenthesis for actually knowing p_5. Provided the dice is 'ideal', then all the p_k equal $1/6$, but were the dice 'charged to reach $p_6 = 0.45$', the rest of the values of p are not yet determined even if they should have a sum equal to $1 - 0.45 = 0.55$ and, for instance, each of them supposed equal to $0.55/5 = 0.11$. Hence, and concerning the probability measures p, there are two different uncertainties being the specification of p the first one, and the lack of exactness on the values of p the second one.

In short, the imprecise character of the predicate P is manifested through $\leq_P \neq =_P$, the uncertainty in the occurrence, or reality, of all the statements or events 'x is P' through the contextually chosen measure p, and the amount of uncertainty on the particular occurrence of each 'x is P' through the indeterminacy associated to compute the particular numerical value of its measuring p (x is P).

VII. Concerning the predicates $P' = $ Not-probable, and $P^a = $ improbable, it should be recalled [21] that it is: $\leq_{P'} \subseteq \leq_P^{-1}$, and $\leq_{Pa} = \leq_P^{-1}$, allowing to state that both are measurable just provided P is so. Their measures, once accepted they are expressible in the forms $\mu_{P'}(x) = N_P(\mu_P(x))$, and $\mu_{Pa}(x) = \mu_P(s_P(x))$, with $N_P : [0, 1] \to [0, 1]$, and $s_P : X \to X$, respectively a negation function and a symmetry, are effectively measures for the corresponding graphs $(X, \leq_{P'})$ and $(X, \leq_{Pa})$. For what respects to their character of 'probabilities', and when the relations are coincidental with the inclusion $\subseteq$ of crisp sets, it should be noticed that μ_{Pa} is a probability since it just permutes the values of the probability μ_P, as it can be easily checked with a finite X, but $\mu_{P'}$ is not since, for instance, $\mu_{P'}(X) = N_P(\mu_P(X)) = N_P(1) = 0$. A scientific study of these predicates is, in fact, an open subject in which, for instance, it could be interesting to find conditions for the coincidence between $\mu_{P'}$ and μ_{Pa} and in correspondence with the often made confusion between the words not-probable and improbable and that, in [20], is done in the very particular case in which X is a Boolean algebra.

3   Reflecting on Fuzzy Versus Probable

3.1

In 1965 Lotfi A. Zadeh founded the theory of Fuzzy sets and systems as a generalization of System theory [28, 29].[1] His intended system was related to pattern classification that was well-known from the field of statistics:

> There are two basic operations: abstraction and generalization, which appear under various guises in most of the schemes employed for classifying patterns into a finite number of categories. [3, p. 1]

Abstraction was to be understood as the problem of identifying a decision function on the basis of a randomly sampled set, and *generalization* referred to the use of the

---

[1] For a detailed history of the theory of Fuzzy sets and its applications see the second author's book [17].

decision function identified during the abstraction process in order to classify the pattern correctly. Although these two operations could be defined mathematically on sets of patterns, Zadeh proposed another way: "a more natural as well as more general framework for dealing with these concepts can be constructed around the notion of a "fuzzy" set - a notion which extends the concept of membership in a set to situations in which there are many, possibly a continuum of, grades of membership." [3, p. 1]

> For example, suppose that we are concerned with devising a test for differentiating between handwritten letters O and D. One approach to this problem would be to give a set of handwritten letters and indicate their grades of membership in the fuzzy sets O and D. On performing abstraction on these samples, one obtains the estimates $\tilde{\mu}_O$ and $\tilde{\mu}_D$ of μ_O and μ_D respectively. Then given a letter x which is not one of the given samples, one can calculate its grades of membership in O and D, and, if O and D have no overlap, classify x in O or D. [29, p. 30]

To get back to Laplace's demon—"who knows it all and rarely tells it all", as Singpurwalla and Booker wrote in [18, p. 872], we refer to these authors' idea to bring this "genie" into "fuzzy" play: They wrote that this "genie is able to classify x with precision, but D is unsure of this classification." The authors of this "scenario" mean by D "a wholeheartedly subjectivist analyst, say D (in honor of [Bruno] de Finetti [1906–1985)])" [18, p. 872].

We do not agree with this view!—We think that there is no part to act for Laplace's demon in the fuzzy game! It is understood that the Laplacian demon would be able to classify the pattern x, but what is supposed to mean: "with precision"?—In our view, Zadeh's pattern classification scenario is in contrast to Laplace's mechanics scenario wherein the demon knows the "true" values of forces and positions of all items. In Zadeh's scenario, there is no "true" classification of pattern x to be an O or a D. A pattern x is a pattern x. Pattern classification means to classify patterns but if there is only the alternative of "O or D" then the answer has to be O or D. Thus, whenever a demon or a human or a machine classifies a pattern x to be an O or to be a D, then this is a subjective decision and even if an "all-knowing" creature does this classification it will still be imprecise, because the given pattern isn't O or D but x.

In their 1995-paper on "Fuzzy Logic and Probability" Petr Hajek, Lluis Godo and Francesc Esteva clarified the basic differences between fuzzy set theory and probability theory by turning the problem to the field of logic:

> Admitting some simplifications, we consider that fuzzy logic is a logic of vague, imprecise notions and propositions that may be more or less true. Fuzzy logic is then a logic of partial degree of truth. On the contrary, probability deals with crisp notions and propositions, propositions that are either true or false; the probability of a proposition is the degree of belief on the truth of that proposition. If we want to consider both as uncertainty degrees we have to stress that they represent very different sorts of uncertainty. (Zimmermann calls them linguistic and stochastic uncertainty, respectively).[2] If we prefer to reserve the word "uncertainty" to refer to degrees of belief, then clearly fuzzy logic does not deal with uncertainty at all. [7, p. 237]

---

[2] The authors referred to the book *Fuzzy Set Theory—and its Applications* of the fuzzy-pioneer Zimmermann [35].

Here, we will use the word "uncertainty" to refer to probabilities and we will use the term imprecision to refer to fuzzy sets.

3.2

In 1968 Zadeh pointed out that in probability theory "an event is a precisely specified collection of points in the sample space. By contrast, in everyday experience one frequently encounters situations in which an "event" is a fuzzy rather than a sharply defined collection of points." In those cases, where uncertainty and imprecision are present in the same problem, he proposed to use probability theory and fuzzy set theory in concert: "...the notions of an event and its probability can be extended in a natural fashion to fuzzy events." In that paper Zadeh gave the following definition of a "fuzzy event" [30, p. 421f]:

Definition 1 Let $(\mathbb{R}^n, \mathbf{A}, P)$ be a probability space in which $\mathbf{A}$ is the σ-field of Borel sets in $\mathbb{R}^n$ and P is a probability measure over $\mathbb{R}^n$. Then, a *fuzzy event* in $\mathbb{R}^n$ is a fuzzy set $\tilde{A}$ in $\mathbb{R}^n$ whose membership function $\mu_{\tilde{A}}$ $\left(\mu_{\tilde{A}} : \mathbb{R}^n \to [0, 1] \right)$, is Borel measurable. The *probability* of a fuzzy event $\tilde{A}$ is defined by the Lebeque-Stieltjes integral:

$$P(A) = \int_{\mathbb{R}^n} \mu_{\tilde{A}}(x) dP = E(\mu_{\tilde{A}}). \tag{1}$$

Thus, [...] the probability of a fuzzy event is the expectation of its membership function."

Singpurwalla and Booker referred to this definition when they proposed that a "wholeheartedly subjectivist analyst" D would have "partial knowledge of the genie's actions" that may be "encapsulated" in D's personal probability, $P_D(x \in \tilde{A})$, that x is in fuzzy set $\tilde{A}$. However, as we argued above, there is still imprecision! Every classification of x—whoever is classifying—is dependent of the definition of concepts— here O or D—which are labelled with letters or words. Therefore it was named "linguistic", by Zimmermann [35] as Hajek et al. mentioned (see above).

In Zadeh's view, probability theory provides a calculus to treat uncertainty about knowledge but *not* a calculus to treat imprecision. In many of his later papers he emphasized that fuzzy set theory should not replace probability theory; moreover he regarded his new theory as an "add on", e.g.:

> ...probability theory is not sufficient for dealing with uncertainty in real world settings. To enhance its effectiveness, probability theory needs an infusion of concepts and techniques drawn from fuzzy logic – especially the concept of linguistic variables and the calculus of fuzzy if-then rules. In the final analysis, probability theory and fuzzy logic are complementary rather than competitive. [32, p. 271]

In the year 2013 Zadeh wrote in an e-mail to the BISC Group[3] that "it is sufficiently important to warrant a brief revisit" concerning Bayesianism[4]:

---

[3]BISC Group is the mailing list of the Berkeley Initiative in Soft Computing (Majordomo@EECS.Berkeley.EDU).

[4]Bayesianism or Bayesian probability theory is named after the English mathematician and Presbyterian Thomas Bayes (1701–1761) but it was popularized by Laplace.

What is Bayesianism? Is it a valid doctrine? There are many misconceptions and many schools of thought. The principal credo of Bayesianism is: Probability is sufficient for dealing with any kind of uncertainty. Call this sufficienism. [34]

As many times before, he quoted the statement of the British statistician Dennis Lindley:

The only satisfactory description of uncertainty is probability. By this I mean that every uncertainty statement must be in the form of a probability; that several uncertainties must be combined using the rules of probability; and that the calculus of probabilities is adequate to handle all situations involving uncertainty ... probability is the only sensible description of uncertainty and is adequate for all problems involving uncertainty. All other methods are inadequate ... anything that can be done with fuzzy logic, belief functions, upper and lower probabilities, or any other alternative to probability can better be done with probability. [14, p. 20]

Then, Zadeh told his readers the following:

I was brought up on probability theory. My first published paper (1949) was entitled "Probability criterion for the design of servomechanisms." My second paper (1950) was entitled "An extension of Wiener's Theory of Prediction." Others followed. I was a sufficienist until 1964, when I converted to insufficienism. My 1965 paper on fuzzy sets reflected this conversion. Since then, I have been involved in many discussions and debates in which I argued that traditional probability theory has intrinsic limitations. My 2002 paper "Toward a perception-based theory of probabilistic reasoning with imprecise probabilities," described my generalization of probability theory–a generalization which involved combining probability theory with possibility theory.[5] [34]

3.3

Under the view of fuzzy sets as quantities this paper holds, there is nothing against the possibility that, in those cases in which the available information can be of a statistical frequency-type, the numerical values $\mu_P(x)$ could be reached 'through' some probability. But this cannot be the general case, as it is argued by some probabilists, and as it is easily shown by very simple examples with fuzzy sets in a finite universe; for instance, in $X = \{x_1, \ldots, x_4\}$, the fuzzy set $\mu = 1/x_1 + 1/x_2 + 0.6/x_3 + 0/x_4$, cannot coincide with any probability since it is $1 + 1 + 0.6 = 2.6 \neq 1$, and if dividing it by 2.6 it is obtained the fuzzy set $\mu^* = \mu/2.6$ that, even with sum equal to 1, the total membership of x_1 and x_2 to μ changes to $1/2.6 = 0.38$ in μ^*. It should be pointed out that is not the same to state that a fuzzy set is but a probability, than to state that their numerical values can be reached either in some statistical way, or each one by a different probability. This happens, for instance, with the former μ and the four probabilities in X, prob_i, $1 \leq i \leq 4$, defined by $\text{prob}_1(x_1) = 1$, and $\text{prob}_1(x_k) = 0$ for $2 \leq k \leq 4$; $\text{prob}_2(x_2) = 1$, and $\text{prob}_2(x_1) = \text{prob}_2(x_3) = \text{prob}_2(x_4) = 0$; $\text{prob}_3(x_3) = 0.6$, and $\text{prob}_3(x_1) + \text{prob}_3(x_2) + \text{prob}_3(x_4) = 1 - 0.6 = 0.4$; $\text{prob}_4(x_4) = 0$, and $\text{prob}_4(x_1) + \text{prob}_4(x_2) + \text{prob}_4(_3) = 1$. In the case X is finite, it does not seem difficult to identify these families of probabilities allowing to define $\mu(x) = \text{prob}_x(x)$. In the finite case, each fuzzy set can be defined by a family

---

[5]For the cited papers see: [16, 27, 33].

of probabilities with, as much, n of them. That is, finite fuzzy sets can be viewed as equivalent to finite families of probabilities. What is not so clear, is how these families of probabilities can be combined between them to reach the intersection, the union, and the pseudo-complement of fuzzy sets. For instance, the fuzzy set $\mu \cdot \mu = \text{prod}(\mu, \mu) = \mu^2 = 1/x_1 + 1/x_2 + 0.36/x_3 + 0/x_4$, can keep prob_1, prob_2, and prob_4, but not prob_3 whose square could not necessarily be, in addition, a probability.

3.4

In principle, fuzzy sets are linked with either the imprecision of their linguistic labels, or with the non-random uncertainty they shown, but probabilities are linked with the random, or repeatable, uncertainty of elements in some particular Ortholattice (Orthomodular or Boolean). If the necessary additive law of probability behaves reasonably with random events, like it is with length, surface, volume, weight, etc., it is much less reasonable with imprecision and non-random uncertainty where some 'interpenetrations' between the events seems to be in their own nature and, hence, the growing of the measure could perfectly be sub-additive, or super-additive as it happens with Zadeh's possibility measures, introduced to deal with non-random uncertainty, whose distributions are fuzzy sets, and that are not coincidental with probabilities.

3.5

Fuzzy sets should be definable in any universe of discourse, either previously structured or not, but crisp-probabilities do require that the universe of the discourse is an Orthomodular lattice or, in particular, a Boolean algebra as in the Kolmogorov's strict case. They are, actually, very strong algebraic structures that, as it is known, cannot hold with all fuzzy sets. Fuzzy sets are definable in universes whatsoever.

Provided the universe X is structured as a Boolean algebra, and two different fuzzy sets in X were defined by $\mu(x) = \text{prob}_1(x)$, and $\sigma(x) = \text{prob}_2(x)$, there is no possibility of having $\mu \leq \sigma$, since $\text{prob}_1(x) \leq \text{prob}_2(x) \Rightarrow \text{prob}_2(x') \leq \text{prob}_1(x')$, for all x' in X, or $\sigma \leq \mu$, and then $\mu = \sigma$. That is, two of these fuzzy sets can only be identical, or not comparable under the usual pointwise ordering of fuzzy sets; this ordering is not suitable for probabilities. In addition, the standard operations with fuzzy sets are not preserved with probabilities; for instance, if $f(x) = \min(\text{prob}_1(x), \text{prob}_2(x))$ is a probability, it will follow $\text{prob}_i \leq f \Rightarrow f = \text{prob}_1 = \text{prob}_2$. Hence, f is not a probability unless both probabilities do coincide, and no algebra of fuzzy sets seems to be definable with this particular kind of fuzzy sets. Notice that in the case of a finite universe with n elements, the set of all fuzzy sets can be viewed as the full unit cube $[0, 1]^n$, but the set of all probabilities is just a polyhedral convex cone in it, and the set of all possibility measures consists in all those n-dimensional points $(x_1, \ldots, x_n)$ with, at least, one component x_i equal to 1. There are more fuzzy sets than possibilities and probabilities. Even more, a probability cannot be self-contradictory: $\text{prob}(x) \leq 1 - \text{prob}(x) \Rightarrow \text{prob}(x) \leq 0.5$, for all x in X, that is absurd since it is, at least, $\text{prob}(X) = 1$.

3.6

If, as it is in almost all cases, in the universe X where the imprecise linguistic labels are applied to, probabilities cannot be defined immediately by, for instance, a lack of a suitable lattice structure, then it is yet possible to study what follows in order that fuzzy sets can be probabilistically expressed. Let us pose the question in mathematical terms [15].

To compute fuzzy sets by probabilities like in the finite case, for all fuzzy set μ_P and each x in X, it should exist a sigma-algebra $\Omega^P(x)$ of subsets in some universe $U^P(x)$, as well as probabilities p^P_x, such that

$$\mu_P(x) = p^P_x(A(x)), \text{ with a crisp set } A(x) \text{ in } \Omega^P(x) \, [**].$$

That is, and departing from X, it is necessary to prove a theorem showing the existence of triplets $(U^P(x), \Omega^P(x), p^P_x)$ for all x in X, verifying [**], and also analyzing the preservation of all the basic laws required to allow an algebra of fuzzy sets with them being pointwise ordered. This is a nice challenge for probabilities, even if the proof is restricted to some particular case, but opening the possibility of characterizing those fuzzy sets than can be probabilistically defined in the form [**], and, also, the negative cases for it. In general, the universes U^P and the sigma-algebras Ω^P will be not coincidental with X like in the former example. The strong case, advocated by some researchers, is that in which all the probabilities p^P_x are coincidental.

For instance, if $\mu_{P \text{ and } Q} = \mu_P \cdot \mu_Q$ (with the operation $\cdot$ intersection), it will require to know a relation between three probabilities defined in different sigma-algebras under which and for all x, the probability associated to P and Q should coincide with some operation between the probabilities associated, respectively, to P and to Q; an expression that, in principle, is not necessarily functionally expressible even if it is so the intersection $\mu_P \cdot \mu_Q$, given, for instance, by a continuous t-norm. It should be analogously clarified what happens with the inclusion of fuzzy sets and the respective sigma-algebras and probabilities, like it is commented in the former paragraph when the universe itself is a sigma-algebra, and for all x in X, all $\Omega^P(x)$ are coincidental with it, and all the p^P_x are just a single probability.

Nevertheless and for what has been said in "3.3", it seems that the general and structural relation [**] asked for, is improbable to exist, and it is very risky to state that all fuzzy sets are either probabilities, or that their numerical values come from probabilities. It is better to cautiously refrain from asserting it before [**], or something similar, is actually proven or disproven.

3.7

For what concerns fuzzy sets in themselves, they usually appear as measures of a linguistic label, or predicate Q acting in a universe of discourse Y, and whose meanings are represented by the corresponding quantities $(Y, \leq_Q, \mu_Q)$. Nevertheless, there is some other kind of problems also generating fuzzy sets as it is, for instance, the following. If two rigid statements a and b are respectively represented by different crisp subsets A and B of X whose respective characteristic or membership functions are f_A and f_B, then there is no a rigid statement c acting in X and represented by the function $F(x) = af_A(x) + bf_B(x)$, with numbers a and b in the real unit interval such that $a + b = 1$. In fact, if $b > 0$ and x is in A, but it is not in B, then $F(x) = a \cdot 1 + b \cdot 0 = a \in (0, 1)$. It analogously happens if x is in B but not in A. That is, the function F belongs to $[0, 1]^{[0,1]} - \{0, 1\}^X$ and, hence, cannot represent a crisp set, but a fuzzy set. Given a function $F : X \to [0, 1]$, finding a predicate C on X such that $F = \mu_C$ is the so-called problem of 'linguistic approximation'; if, in praxis and usually by comparison with previously known fuzzy sets, it is often found one of such predicates even without the safety that its representation by a quantity verifies $F = \mu_C$, the problem is actually open, as it is with crisp sets, and possibly it has neither a single solution, nor it exists a systematic method for specifying C. It should be remembered that the axiom of specification states that a precise predicate specifies a single subset, but not that to every subset it corresponds a single predicate naming it.

4 Zadeh's (Numerical) Crisp-Probability of Fuzzy Events

4.1

When a statement involving imprecise predicates can be represented by means of a fuzzy set, and as it happens when a statement involving precise ones is represented by a crisp set (for instance, the statements 'obtaining odd' in throwing a dice is represented by the set $\{1, 3, 5\}$, 'between 4 and 7' concerning a variable is represented by the interval $[4, 7]$ in the real line, etc.), its probability should be given through the corresponding representation either in crisp, fuzzy or both kind of terms. As it is always in Science, there is no way of establishing a formal theory without a suitable setting representing what the theory involves. Hence, it is important to define what can be understood by the probability of a 'fuzzy event', a fuzzy set translating a linguistic statement; a probability that, according to language's use, can be numerical (the probability that John is short is 0.85), or linguistic (the probability that John is short is high), in which case its values will be, at its turn, fuzzy sets (μ_{short}, for instance). That is, the numerical probability of fuzzy events, that we will shorten respectively as the fuzzy-crisp or the fuzzy-fuzzy probability depending on where the probability values range, should be respectively represented by crisp numbers or by fuzzy numbers. In any case, the probability of fuzzy events in $[0, 1]^X$, deserves to be posed by either ranging in $[0, 1]$, or in $[0, 1]^{[0,1]}$. Let us first consider the fuzzy-crisp probability [15] for which, copying the previously presented tow axioms, is a mapping $p : [0, 1]^{[0,1]} \to [0, 1]$, such that:

1. Normalization property: $p(\mu_1) = 1$, with μ_1 the function constantly equal to 1, that is, the membership function of the full crisp set $[0, 1]$.
2. Additive property: If either $\mu \le \sigma'$, or $\mu \cdot \sigma = \mu_0$ (the function constantly equal to 0, that is, the membership function of the empty set $\emptyset$), then $p(\mu + \sigma) = p(\mu) + p(\sigma)$.

To being actually a measure of fuzzy events, such function should necessarily verify the property: $\mu \leq \sigma \Rightarrow p(\mu) \leq p(\sigma)$, and to reach it from 2, it is necessary to work in a suitable algebra of fuzzy sets expressing the pseudo-complement ('), the intersection ($\cdot$), and the union (+) of fuzzy sets, as well as to know when contradiction implies, or is equivalent, to incompatibility.

In what follows only standard theories of fuzzy sets will be taken into account, that is, those functionally expressible in which the intersection ($\cdot$) is given by a continuous t-norm T ($\mu \cdot \sigma = T \circ (\mu \times \sigma)$), the union (+) by a continuous t-conorm S ($\mu + \sigma = S \circ (\mu \times \sigma)$), and the pseudo-complement (') by a strong negation N ($\mu' = N \circ \mu$).

4.2

Concerning contradiction, two fuzzy sets μ, and σ, are contradictory—if and only if—$\mu \leq \sigma' = N \circ \sigma \leq \phi^{-1}(1 - \phi \circ \sigma) \Leftrightarrow \phi(\mu(x)) \leq 1 - \phi(\sigma(x)) \Leftrightarrow \phi(\mu(x)) + \phi(\sigma(x)) \leq 1$, for all x, where ϕ is an order-automorphism of the unit interval, and provided it is $N \leq N_\phi$. It should be recalled [26] that it is $N_\phi = \phi^{-1} \circ (1 - \phi)$, the strong negation generated by ϕ.

For what concerns incompatibility, the functional equation able to give $\mu \cdot \sigma = \mu_0$, that is, $T(\mu(x), \sigma(x)) = 0$, for all x in X ($T(a, b) = 0$, in the unknown T) should be solved for all the continuous t-norms. Since the only t-norms with zero divisors are those in the Łukasiewicz family $T = W_\theta = \theta^{-1} \circ W \circ (\theta \times \theta)$, a first result follows immediately: There is incompatibility between μ and σ under a Łukasiewicz's t-norm W_θ if and only if $\max (0, \theta(\mu(x)) + \theta(\sigma(x))) = 0 \Leftrightarrow \theta(\mu(x)) + \theta(\sigma(x)) = 0 \Leftrightarrow \mu(x) = \sigma(x) = 0$, for all x, that is, $\mu = \sigma = \mu_0$, in which case it is also contradiction between both fuzzy sets. Only with T in the Łukasiewicz family there is equivalence between contradiction and incompatibility.

For what concerns to $T = \min$ and $T = \text{prod}_\theta$, since $T(a, b) = 0 \Leftrightarrow a = 0$, or $b = 0$, it follows $T(\mu(x), \sigma(x)) = 0 \Leftrightarrow$ Either $\mu(x) = 0$, or $\sigma(x) = 0$, for all x. In both cases, $\mu \leq \sigma'$, and there is also contradiction.

To summarize: Without considering ordinal-sums as t-norms, in all cases in which the t-norm belongs to the families of min, prod, and W, incompatibility between μ and σ is obtained whenever, at each point x, either $\mu(x) = 0$, or $\sigma(x) = 0$, or both. In all cases, incompatibility implies contradiction, but, in general, both concepts are not equivalent except if the t-norm belongs to Łukasiewicz family, and then the situation is similar to that in Boolean algebras, but it is not like what it happens with the quantum probability in Orthomodular lattices, where contradiction implies incompatibility. Consequently, to define a numerical probability for fuzzy sets, it should be chosen how to define its additive property, either by means of contradiction, or incompatibility.

Concerning the second main property the measure p should enjoy, that is, that from $\mu \leq \sigma$ follows $p(\mu) \leq p(\sigma)$, it is necessary to count with a deduction process analogous to that in Boolean algebras, that is, for instance, the validity of the functional equation $\mu = \mu \cdot \sigma + \mu \cdot \sigma' \Leftrightarrow \mu(x) = S(T(\mu(x), \sigma(x)), T(\mu(x), N(\sigma(x)))) \Leftrightarrow S = W_\phi^*, T = W_\phi$, and $N = N_\phi$, to actually knowing that the additive law of p implies its numerical values actually grow when the events 'grow' in the pointwise order.

In these cases, and if there is in addition incompatibility (and hence contradiction), from the last equation it does not follow $\mu \leq \sigma \Rightarrow \mu = \mu \cdot \sigma$, which only holds provided either $\mu(x) = 0$, or $\sigma(x) = 1$, and, hence, the same way to reach the conclusion does not hold, and only holds the orthomodular law $\mu \leq \sigma \Leftrightarrow \sigma = \mu + \sigma \cdot \mu'$, always valid in the sense $\sigma = \mu + \sigma \cdot \mu' \Rightarrow \mu \leq \sigma$, but that reciprocally only holds in the same case that the law of perfect repartition, that is, with $S = W_\phi^*$, $T = W_\phi$, and $N = N_\phi$. Hence, it seems that a theory of probability for fuzzy sets can only be developed in a form very close to the classical theory provided the algebra is given by these S, T, and N.

4.3

Nevertheless, Zadeh reached a way to escape from this general view [30], by defining a probability for fuzzy sets that avoids such problem, although it shows a shortcoming with conditionality. Zadeh took the universe of discourse X as $\mathbb{R}^n$, that is good enough for many applications. In it he considered those fuzzy sets $\mu : \mathbb{R}^n \to [0, 1]$ that are Borel measurable, and defines $p(\mu)$ (see (1) in "3.2") as the 'formal expectation' of μ:

$$p(\mu) = \int_{\mathbb{R}^n} \mu = E(\mu),$$

the Lebesgue-Stieltjes integral of μ, with which it immediately follows that $\mu \leq \sigma$ implies $p(\mu) \leq p(\sigma)$, regardless of the chosen algebra of fuzzy sets. What Zadeh implicitly supposed is that the function μ acts as a random variable. Since it is also obvious that $p(\mathbb{R}^n) = 1$, and $p(\emptyset) = 0$, p allows to measure the amount of P = probable shown by the Borel measurable fuzzy sets in $\mathbb{R}^n$, and provided there is coincidence between the pointwise order of fuzzy sets and the binary relation $\leq_P$, 'less P than'. For all the continuous t-norms and t-conorms verifying $T(a, b) + S(a, b) = a + b$ (Frank's family), it obviously follows,

$$p(\mu \cdot \sigma) + p(\mu + \sigma) = p(\mu) + p(\sigma),$$

and thus it also follows the additive law:

$$\mu \cdot \sigma = \mu_0 \Rightarrow p(\mu + \sigma) = p(\mu) + p(\sigma),$$

with which p can be seen like a probability in the classic Kolmogorov's sense for Boolean algebras. Hence, in the case in which the statements S on a universe of discourse X can be represented by fuzzy sets μ_S in $\mathbb{R}^n$, it can be said that the measure of how probable they are is $p(\mu_S)$, provided we are in the setting of a standard theory of fuzzy sets whose t-norm and t-conorm belong to the Frank's family.

There is, notwithstanding, a shortcoming with this definition [15], since the corresponding conditional probability cannot be taken into account for all the t-norms in the Frank's family, but only with $T = \min$. In fact, and for instance, with $T = \text{prod}$, the typical definition

$$p(\mu/\sigma) := p(\mu \cdot \sigma)/p(\sigma), \text{ with } p(\sigma) > 0,$$

is not a probability among the fuzzy sets in the 'diaphragmed' subset whose elements have the form $\mu \cdot \sigma$, since $p(\sigma/\sigma) = p(\sigma^2/\sigma)$ is not always equal to one. Neither is it with $T = W$, since $p(\sigma/\sigma) = p$ (max $(0, 2\sigma - 1))/p(\sigma) = 1$ will not always hold if σ is different from μ_1, and is 0 provided $\sigma \le \mu_{1/2}$. Obviously, with $T = \min$ and since $\sigma \cdot \sigma = \min(\sigma, \sigma) = \sigma$, it holds $p (\sigma/\sigma) = 1$, and since the corresponding t-conorm in the Frank's family is $S = \max$, it is easy to prove $p(\mu + \pi/\sigma) + p(\mu \cdot \pi/\sigma) = p(\mu/\sigma) + p(\pi/\sigma)$, and it results that $p(\cdot/\sigma)$ is a probability in the 'diaphragmed' set. Only with $T = \min$ and $S = \max$, the conditional probability is actually a probability. Hence, it seems that such pair of connectives are the only suitable ones for defining a Kolmogorov's probability with fuzzy sets.

Notice that only with $T = \min$, and $S = \max$, is $[0, 1]^X/\sigma = \{\mu \cdot \sigma; \mu \in [0, 1]^X\}$ closed by intersection and union: $\min(\mu_1 \cdot \sigma, \mu_2 \cdot \sigma)$ in each point x is equal to one of both and hence it is in $[0, 1]^X/\sigma$, and analogously with max. Only in this case they belong to the 'diaphragmed' set. Hence, the set of fuzzy sets $[0, 1]^X/\sigma$, with $p(\sigma) > 0$, and the couple (min, max) is also an Standard algebra of fuzzy sets.

Since, with $T = \text{prod } (\cdot)$, it is $T(\mu_1 \cdot \sigma, \mu_2 \cdot \sigma) = (\mu_1 \cdot \mu_2 \cdot \sigma) \cdot \sigma$, and also with its dual t-conorm $S = \text{sum-prod}$, $S(\mu_1 \cdot \sigma, \mu_2 \cdot \sigma) = (\mu_1 + \mu_2 - \mu_1 \cdot \mu_2 \cdot \sigma) \cdot \sigma$, $[0, 1]^X\sigma$ is also a standard algebra of fuzzy sets with the couple (prod, prod*). Nevertheless, it is not with $T = W$ and $S = W^*$, as it can be easily proven. Hence, with the couple (prod, prod*), p is a probability whose corresponding conditional probability is not a probability.

Remarks

I. A probability for fuzzy events in $\mathbb{R}^n$ can be defined at any standard algebra of fuzzy sets $([0, 1]^{\mathbb{R}^n}, T, S, N)$ with T and S in the Frank's family, but, provided a conditional probability is required, then it only can be obtained with $T = \min$, and $S = \max$. Hence, to count with a conditional probability it is necessary to extent the pair (T, S) to the quartet $(T, S, \min, \max)$ that, only if $T = \min$ and $S = \max$ is reduced to the pair (min, max). This is the only case fully mimicking the classical Kolmogorov one.

II. It should be noticed that with quantum probability in Orthomodular lattices, there is not a generally accepted way of introducing conditional probability [5].

III. The same results are obtained provided the fuzzy sets are taken to be [15] Riemann integrable instead of Lebesgue integrable in $\mathbb{R}^n$. Even if this is somewhat restrictive it is sufficient for most of the practical cases. For instance, if $X = [0, 10]$, $\mu(x) = x/10$ and $\sigma(x) = 1 - x/10$, it is $p(\mu) = 1/100 \int_0^{10} x dx = 0.5$, and $p(B/A) = p(\sigma \cdot \mu)/0.5 = p(x/10 - x^2/100)/10 = 0.8333 \dots$ When μ is a crisp set, for instance the subinterval $A = [1, 3]$ of $[0, 10]$, the probability reduces to $p(A) = 1/10 \int_1^3 dx = 0.2$, only coincidental with the classical probability when this is related to a uniform distribution.

5 Some Comments on Fuzzy-Fuzzy Probabilities

5.1

To pose the concept of a 'fuzzy-fuzzy probability' [15], that is, when the events are fuzzy sets, the probability is expressed linguistically, and that can be exemplified, for instance, by "The probability that 'John is tall' is high", represented by:

$$\text{Prob (John is tall)} = \text{High, or, in fuzzy terms, } p(\mu_{tall}(John)) = \mu_{high},$$

with μ_{tall} in $[0, 1]^X$, and μ_{high} in $[0, 1]^{[0,1]}$, it is required to count with a suitable algebra in $F(X) = [0, 1]^X$, and another one in $I = [0, 1]^{[0,1]}$, since it should be $p : F(X) \rightarrow I$. In the same vein that before and with the aim of having a general theory like it happens in classic crisp-probability with sets, it will be supposed that the binary relation $\leq_p$ coincides with the pointwise ordering between fuzzy sets. It will be analogously supposed that I is the poset given by the pointwise order and whose minimum and maximum are, respectively, μ_0 and μ_1.

If $X \subseteq \mathbb{R}^n$, to generalize Zadeh's fuzzy-crisp probabilities defined in "4.1", it could be presumed that if the values of p are 'numbers' A_r (fuzzy sets in $[0, 1]$ constantly equal to $r \in [0, 1]$), then p can be analogously defined but with values in I. Hence, a fuzzy-fuzzy probability is a mapping $p : F(X) \rightarrow I$, such that:

- $p(A_1) = p(X) = \mu_1$
- $\mu \cdot \sigma = \mu_0 \Rightarrow p(\mu + \sigma) = p(\mu) + p(\sigma)$
- If $X = R^n$, and the values of p are numbers A_r in I, p is a Zadeh's fuzzy-crisp probability.

The triplet of operations $(\cdot, +, ')$, both in $F(X)$ and in I, should be chosen in such a way that:

- $p(\mu') = \mu_1 - p(A) \Leftrightarrow p(A) + p(A') = \mu_1$
- $\mu \leq \sigma \Rightarrow p(\mu) \leq p(\sigma)$
- $p(\mu_0) = \mu_0$,

for properly calling p a probability.

5.2

For what concerns the values of p in I, it could be sometimes suitable to take them as 'fuzzy numbers' N_r, that is, fuzzy sets in $[0, 1]^{[0,1]}$ whose membership functions are of the form [26],

- $N_r(x) = 0$, if $0 \leq \times \leq r - \epsilon$, or $r + \epsilon \leq \times \leq 1$,
- $N_r(x) = L(x)$, if $r - \epsilon \leq \times \leq r$,
- $N_r(x) = R(x)$, if $r \leq \times \leq r + \epsilon$,

provided $\epsilon > 0$, and with functions $L : [r - \epsilon, r] \rightarrow [0, 1]$, and $R : [r, r + \epsilon] \rightarrow [0, 1]$, verifying:

- L is strictly non-decreasing between $r - \epsilon$ and r, with $L(r - \epsilon) = 0$, and $L(r) = 1$,
- R is strictly decreasing between $r + \epsilon$ and r, with $R(r) = 1$, and $R(r + \epsilon) = 0$,

Sometimes, L and R are supposed to be linear, that is, $L(x) = (r - \epsilon - x)/r$, and $R(x) = (r + \epsilon - x)/e$.

In such cases, the set of fuzzy numbers $\{N_r; r \in [0, 1]\}$ could need to be endowed with a suitable partial ordering with extreme elements of the types:

- $N_0(x) = R(x)$, in a (right) neighborhood of 0, and
  $N_0(x) = 0$ in its complement in $[0, 1]$,
- $N_1(x) = L(x)$, in a (left) neighborhood of 1, and
  $N_1(x) = 0$ in its complement in $[0, 1]$.

Notice that N_r can represent, for instance, 'around r', and that to employ these fuzzy numbers corresponds to problems like 'the probability that John is rich is around 0, 6'. In these cases, not only the ordering should be adapted to fuzzy numbers, but also the operations between them should be those of Fuzzy Arithmetic generalizing the classical arithmetic operations through the Fuzzy Logic's Extension Principle [26], and to keep the classical arithmetic when N_r can coincide with r.

5.3

It is important to carefully establish the operations in $F(X)$ and in I. In general, and in the same vein as when representing a dynamical system whose behaviour is linguistically described, they may not be universal, and different cases may require different triplets of them, but justifications for using a particular triplet should be provided at each case; modeling requires a careful design of the involved fuzzy terms [22]. The effectiveness of these operations will depend on the used types of fuzzy sets as values of the fuzzy-fuzzy probability. Therefore and to fully establish a mathematically rigorous and fertile theory for fuzzy-fuzzy probabilities without going far from Kolmogorov's ideas, various components of fuzzy systems must be properly designed. Of course, this is still a serious theoretical open problem that concerns to link what is in language in the form 'the probability of a is b' with good enough mathematical representations of the form prob $(\alpha) = \beta$, where the fuzzy set α represents statement a, the fuzzy set β statement b, and prob is also a suitable representation of the word 'probability'.

In fact, fuzzy-fuzzy probability is, as a mathematical subject, an open one that is waiting for its development. A development that should be grounded on practical cases previously studied, like classical probability was grounded on the study of games of chance, and the errors in computation. In this case, it seems necessary to know more on what happens in language with the qualitative interpretation of the predicate probable with imprecise statements. The development of fuzzy logic should evolve towards a science of imprecision and non-random uncertainty.

6 Conclusions

6.1

From its very inception fuzzy sets kept a difficult and sometimes troublesome relation with probability [31], often coming from a not clear enough view of the respective grounds. Fuzzy sets mainly deal with linguistic imprecision and non-repetitive uncertainty of predicates and in this view are nothing else, but nothing less, than measures of their meaning, whilst probability just measures the uncertainty of the outcomes in a given experiment, whose realization is presumed to be repeatable in a large number of times, and always under the same conditions. The 'physical' character of the outcomes in a random experiment, those typically considered by probability theory, like they are the extractions of balls in a urn, makes that probabilities grow, without no doubt, additively, that is they enjoy the additive law, from which it can be deduced that probabilities grow with the growing of events. Both probabilities and fuzzy sets measure information, but different types of information.

To compare probability with fuzzy sets, it should be taken into account what they apply to, which properties define each entity, and of course, what they respectively measure. To define a fuzzy set it is not necessary to apply its linguistic label to a universe previously endowed with any algebraic structure, but as it has been seen, no probability can be defined without counting with a previous, and strong, structure among the elements to which it is applied.

6.2

In the theory of fuzzy sets, there is a, sometimes not explicitly expressed, principle forcing to keep what classically happens with crisp sets, that is, when the membership function only takes the values 0 and 1. Such principle is that of 'Preservation of the Classical Case' [26], and comes from the necessity of jointly working with imprecise and precise concepts. For instance, if the universe is the unit interval $[0, 1]$, the mapping $* : [0, 1] \rightarrow [0, 1]$, defined by $\mu^*(x) = 1 - \mu(1 - x)$, that verifies all the properties of a negation between fuzzy sets, cannot be taken as a pseudo-complement for fuzzy sets since although the image by it of a crisp set is a crisp set, it is not always its crisp complement. The mapping $*$ does not preserve the classical case, and hence, it cannot be accepted as a pseudo-complement for any algebra of fuzzy sets as it is, for instance, the mapping $\mu'(x) = 1 - \mu(x)$, for x in any universe X.

Analogously, and when the events are expressed by fuzzy sets, any definition of their probability, either valued in $[0, 1]$, or in $[0, 1]^{[0,1]}$, should keep the universally accepted basic laws of a classical crisp-crisp probability. Either fuzzy-crisp, or fuzzy-fuzzy probability, should preserve the classical case since, in praxis, there are cases in which flexible and rigid linguistic labels coexist in the linguistic description of the same problem. It does not happen with Zadeh's fuzzy-crisp probability; for instance, in $X = [0, 1]$, it is $p(\mu_r) = \int_0^1 r dx = r$, contrarily to the fact that the probability of a point under a continuous distribution is always nul.

6.3

Although the existing lack information on the subject, it can be conjectured that the word 'probable', or a synonymous of it, were used before mathematicians begin with the study of 'how much probable' is to obtain an outcome in some random experiments, like those (finite) with carts or dices, and that George Pólya did identify with the extraction of balls from urns. In those experiments, the universe of the outcomes that can be expected is finite and, consequently, it has sense to identify

'statement a is less probable than statement b' with '$A \subseteq B$',

since the number of elements in B is greater or equal to that in A, and provided A and B are the subsets representing the outcomes that correspond to a and b, respectively. Nevertheless, when the statements are not referring to a perfectly isolable situation, like they are, for instance, a question on the beauty of a work of art that can deserve the answer 'is with high probability that it is beautiful', or 'the probability that John is tall is no less than 0.8', or 'it is with a low probability than a few black balls can be extracted from a urn with much more blue tan black balls', etc., it is not clear to what the relation $\leq_{probable}$ can be identified with the goal of establishing a coherent mathematical theory like, it is the classical one as it was described by Kolmogorov's axiomatics. Notwithstanding, if it can be stated that the three properties a measure should necessarily enjoy must be satisfied, it is not so clear what should happen with both the additive property and the probability of the negation. These are two basic laws of the crisp-crisp probability not presenting a real problem in the case of the fuzzy-crisp probabilities introduced by Zadeh, provided it is $X = \mathbb{R}^n$, with an obvious problem existing if the universe of discourse X is not representable as a part of some $\mathbb{R}^n$. When trying to model by fuzzy sets the linguistic or fuzzy-fuzzy probability, it should be based in clear reasons that $\leq_{probable}$ can be identified with the pointwise ordering between fuzzy sets.

6.4

To finish, the mathematical study of fuzzy-fuzzy probability as a continuation of the classical theory of (crisp) probability still shows problems that should be clarified before talking of a theory of fuzzy probability. For what concerns the fuzzy-crisp probabilities in, at least, the definition of Zadeh, it seems that the algebras of fuzzy sets, like the triplets giving the Standard ones are not enough, but that families of more than three connectives are more suitable. Anyway, what it seems still lacking is an experimental study concerning the use in language of the linguistic label 'probable'.

Acknowledgments First author work is partially supported by the Foundation for the Advancement of Soft Computing (Asturias, Spain).

References

1. Alsina, C., Trillas, E., Pradera, A.: On a class of fuzzy set theories. In: Proceedings FUZZ-IEEE, pp. 1–5 (2007)
2. Aristotle, Metaphysics, (I 1, 980 a 21), written 350 B.C., transl. by W. D. Ross. http://classics.mit.edu/Aristotle/metaphysics.1.i.html
3. Bellman, R.E., Kalaba, R., Zadeh, L.A.: Abstraction and pattern classification, Memo RM-4307-PR, Santa Monica, CA, The RAND Corporation, Oct. 1964. Later in: J. Math. Anal. Appl. 13, 1–7 (1996)
4. Birkhoff, G.: Lattice Theory. AMS Pubs, Providence (1967)
5. Bodiou, G.: Théorie dialectique des probabilités (englobant leurs calculs classique et quantique). Gauthier-Villars, Paris (1965)
6. Efron, B.: Bayesians. Frequentists Sci. J. Am. Stat. Assoc. 100, 409 (2005)
7. Hajek, P., Godo, L., Esteva, F.: Fuzzy logic and probability. In: Proceedings of the Eleventh Annual Conference on Uncertainty in Artificial Intelligence (UAI-95), San Francisco, CA, Morgan Kaufmann, pp. 237–244 (1995)
8. Kappos, D.A.: Probability Algebras and Stochastic Spaces. Academic Press, New York (1969)
9. Kolmogorov, A.N.: Grundbegriffe der Wahrscheinlichkeitsrechnung (in German). Julius Springer, Berlin (1933). Engl. transl.: [10]
10. Kolmogorov, A.N.: Foundations of the Theory of Probability. Chelsea, New York (1956)
11. Laplace, P.S.: Mémoire sur les probabilités, Mém. Acad. R. Sci. Paris, 1778, 227–332 (1781). Oeuvres 9, pp. 383–485. Engl. transl. by R. Pulskamp: http://www.cs.xu.edu/math/Sources/Laplace/index.html
12. Laplace, P.S.: Essai philosophique sur les probabilits, Paris: Courcier, : Cited after the English edition: A Philosophical Essay on Probabilities (Classic Reprint), p. 1902. Chappman and Hall, Wiley, London (1814)
13. Leibniz, G.W.: Von dem Verhängnisse. In: Hauptschriften zur Grundlegung der Philosophie, vol II, pp. 129–134. Ernst Cassirer, Leipzig (1906)
14. Lindley, D.V.: The probability approach to the treatment of uncertainty in artificial intelligence and expert systems. Stat. Sci. 2, 17–24 (1987)
15. Nakama, T., Trillas, E., García-Honrado, I.: Axiomatics investigation of fuzzy probabilities. In: Seising, R., Sanz, V. (eds.) Soft Computing in Humanities and Social Sciences, pp. 125–140. Springer, Berlin (2012)
16. Ragazzini, J.R., Zadeh, L.A.: Probability criterion for the design of servomechanisms. J. Appl. Phys. 20, 141–144 (1949)
17. Seising, R.: The Fuzzification of Systems. The Origins of Fuzzy Set Theory and Its Initial Applications—Its Development to the 1970s. Springer, Berlin (2007)
18. Singpurwalla, N., Booker, J.M.: Membership functions and probability measures of fuzzy sets. J. Am. Stat. Assoc. 99(467), 867–876 (2004)
19. Trillas, E., Alsina, C., Termini, S.: Didactical note: probabilistic conditionality in a boolean algebra. Mathware Soft Comput. 3(1–2), 149–157 (1996)
20. Trillas, E.: On the words not-probable and improbable. In: Proceedings IPMU'2000, vol II, pp. 780–783. Madrid (2000)
21. Trillas, E.: On a Model for the Meaning of Predicates. In: Seising, R. (ed.) Views of Fuzzy Sets and Systems from Different Perspectives, pp. 175–205. Springer, Berlin (2009)
22. Trillas, E., Guadarrama, S.: Fuzzy representations need a careful design. Int. J. Gen. Syst. 39(3), 329–346 (2010)
23. Trillas, E., Urtubey, L.: Towards the dissolution of the Sorites Paradox. Appl. Soft Comput. 11, 1506–1510 (2011)
24. Trillas, E.: Some uncertain reflections on uncertainty. Arch. Philos. Hist. Soft Comput. 1, 1–12 (2013)
25. Trillas, E.: How science domesticates concepts? Arch. Philos. Hist. Soft Comput. 1, 1–13 (2014)

26. Trillas, E., Eciolaza, L.: Fuzzy Logic. An Introductory Course for Engineering Students. Springer, Berlin (2015)
27. Zadeh, L.A., Ragazzini, J.R.: An extension of Wiener's theory of prediction. J. Appl. Phys. **21**, 645–655 (1950)
28. Zadeh, L.A.: Fuzzy sets. Inf. Control **8**, 338–353 (1965)
29. Zadeh, L.A.: Fuzzy sets and systems. In: Fox, J. (ed.) System Theory. Microwave Res. Inst. Symp., Series XV, pp. 29–37. Polytechnic Press, New York (1965)
30. Zadeh, L.A.: Probability measures of fuzzy events. J. Math. Anal. Appl. **23**(2), 421–427 (1968)
31. Zadeh, L.A.: Fuzzy probabilities. Inf. Process. Manage. **20**(3), 363–372 (1984)
32. Zadeh, L.A.: Discussion: probability theory and fuzzy logic are complementary rather than competitive. Technometrics **37**(3), 271–275 (1995)
33. Zadeh, L.A.: Toward a perception-based theory of probabilistic reasoning with imprecise probabilities. J. Stat. Plann. Infer. **105**, 233–264 (2002)
34. Zadeh, L.A.: Bayesianism—A brief revisit, e-mail to BISC Group, 2013/07/31
35. Zimmermann, H.-J.: Fuzzy Set Theory—and its Applications, Boston, Dordrecht. Kluver Academic Publ, London (1991)

Organizing Families of Aggregation Operators into a Cube of Opposition

Didier Dubois, Henri Prade and Agnès Rico

Abstract The cube of opposition is a structure that extends the traditional square of opposition originally introduced by Ancient Greek logicians in relation with the study of syllogisms. This structure, which relates formal expressions, has been recently generalized to non Boolean, graded statements. In this paper, it is shown that the cube of opposition applies to well-known families of idempotent, monotonically increasing aggregation operations, used in multiple criteria decision making, which qualitatively or quantitatively provide evaluations between the minimum and the maximum of the aggregated quantities. This covers weighted minimum and maximum, and more generally Sugeno integrals on the qualitative side, and Choquet integrals, with the important particular case of Ordered Weighted Averages, on the quantitative side. The main appeal of the cube of opposition is its capability to display the various possible aggregation attitudes in a given setting and to show their complementarity.

1 Introduction

The application of fuzzy sets [1] to multiple criteria decision making [2] has led to the continued blossoming of a vast amount of studies on different classes of aggregation operators for combining membership grades. This includes in particular triangular norms and co-norms [3] on the one hand, and Sugeno and Choquet integrals [4, 5] on the other hand. Ronald Yager, in his vast amount of important contributions to fuzzy set theory on many different topics, has been especially at the forefront of

D. Dubois (✉) · H. Prade
IRIT, CNRS & University of Toulouse, 31062 Toulouse Cedex 9, France
e-mail: dubois@irit.fr

H. Prade
e-mail: prade@irit.fr

A. Rico
ERIC, Université Claude Bernard Lyon 1, 69100 Villeurbanne, France
e-mail: agnes.rico@univ-lyon1.fr

© Springer International Publishing Switzerland 2017
J. Kacprzyk et al. (eds.), *Granular, Soft and Fuzzy Approaches for Intelligent Systems*, Studies in Fuzziness and Soft Computing 344,
DOI 10.1007/978-3-319-40314-4_2

creativeness regarding aggregation operators, with in particular the introduction of a noticeable family of triangular norms and co-norms [6], of uninorms [7], and of Ordered Weighted Averages (OWA) [8–10].

Sugeno and Choquet integrals are well-known families of idempotent, monotonically increasing aggregation operators, used in multiple criteria decision making, with a qualitative and a quantitative flavor respectively. They include weighted minimum and maximum, and weighted average respectively, as particular cases, and provide evaluations lying between the minimum and the maximum of the aggregated quantities. In such a context, the gradual properties corresponding to the criteria to fulfill are supposed to be positive, i.e., the global evaluation increases with the partial ratings. But some decisions or alternatives can be found acceptable because they do not satisfy some (undesirable) properties. So, we also need to consider negative properties, the global evaluation of which increases when the partial ratings decreases. This reversed integral is a variant of Sugeno integrals, called desintegrals [11, 12]. Their definition is based on a decreasing set function called anti-capacity. Then, a pair of evaluations made of a Sugeno integral and a reversed Sugeno integral is useful to describe acceptable alternatives in terms of properties they must have and of properties they must avoid.

Besides, we can distinguish the optimistic part and the pessimistic part of any capacity [13]. It has been recently indicated that Sugeno integrals associated to these capacities and their associated desintegrals form a cube of opposition [14], the integrals being present on the front facet and the desintegrals on the back facet of the cube (each of these two facets fit with the traditional views of squares of opposition [15]). As this cube exhausts all the evaluation options, the different Sugeno integrals and desintegrals present on the cube are instrumental in the selection process of acceptable choices. We show in this paper that a similar cube of opposition exists for Choquet integrals, which can then be particularized for OWA operators.

The paper is organized as follows. Section 2 provides a brief reminder on the square of opposition, and introduces the cube of opposition and its graded extension in a multiple criteria aggregation perspective. Section 3 restates the cube of opposition for Sugeno integrals and desintegrals. Section 4 presents the cube for Choquet integrals and then for OWA operators, and discusses the different aggregation attitudes and their relations.

2   Background and Notations

We first recall the traditional square of opposition originally introduced by Ancient Greek logicians in relation with the study of syllogisms. This square relates universally and existentially quantified statements. Then its extension into a cube of opposition is presented, together with its graded version, in a qualitative multiple criteria aggregation perspective.

Fig. 1 Square of opposition

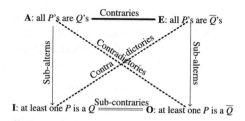

2.1 The Square and Cube of Opposition

The traditional square of opposition [15] is built with universally and existentially quantified statements in the following way. Consider a statement (**A**) of the form "all P's are Q's", which is negated by the statement (**O**) "at least one P is not a Q", together with the statement (**E**) "no P is a Q", which is clearly in even stronger opposition to the first statement (**A**). These three statements, together with the negation of the last statement, namely (**I**) "at least one P is a Q" can be displayed on a square whose vertices are traditionally denoted by the letters **A, I** (affirmative half) and **E, O** (negative half), as pictured in Fig. 1 (where $\overline{Q}$ stands for "not Q").

As can be checked, noticeable relations hold in the square:

(i) **A** and **O** (resp. **E** and **I**) are the negation of each other;
(ii) **A** entails **I**, and **E** entails **O** (it is assumed that there is at least one P, to avoid existential import problems);
(iii) together **A** and **E** cannot be true, but may be false;
(iv) together **I** and **O** cannot be false, but may be true.

Changing P into $\neg P$, and Q in $\neg Q$ leads to another similar square of opposition **aeoi**, where we also assume that the set of "not-P's" is non-empty. Then the 8 statements, **A, I, E, O, a, i, e, o** may be organized in what may be called a *cube of opposition* as in Fig. 2.

This cube first appeared in [16] in a renewed discussion of syllogisms, and was reintroduced recently in an information theoretic perspective [17]. The structural properties of the cube are:

Fig. 2 The cube of opposition for quantified statements

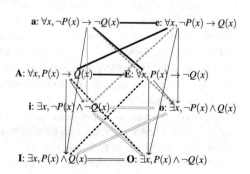

- **AEOI** and **aeoi** are squares of opposition,
- **A** and **e**; **a** and **E** cannot be true together,
- **I** and **o**; **i** and **O** cannot be false together,
- **A** entails **i**, **E** entails **o**, **a** entails **I**, **e** entails **O**.

In the cube, if we also assume that the sets of "Q's" and "not-Q's" are non-empty, then the thick non-directed segments relate contraries, the double thin non-directed segments sub-contraries, the diagonal dotted non-directed lines contradictories, and the vertical uni-directed segments point to subalterns, and express entailments.

Stated in set-theoretic notation, **A**, **I**, **E**, **O**, **a**, **i**, **e**, **o**, respectively mean $P \subseteq Q$, $P \cap Q \neq \emptyset$, $P \subseteq \overline{Q}$, $P \cap \overline{Q} \neq \emptyset$, $\overline{P} \subseteq \overline{Q}$, $\overline{P} \cap \overline{Q} \neq \emptyset$, $\overline{P} \subseteq Q$, $\overline{P} \cap Q \neq \emptyset$. In order to satisfy the four conditions of a square of opposition for the front and the back facets, we need $P \neq \emptyset$ and $\overline{P} \neq \emptyset$. In order to have the inclusions indicated by the diagonal arrows in the side facets, we need $Q \neq \emptyset$ and $\overline{Q} \neq \emptyset$, as further normalization conditions.

Suppose P denotes a set of important properties, Q a set of satisfied properties (for a considered object). Vertices **A**, **I**, **a**, **i** correspond respectively to 4 different cases:

(i) all important properties are satisfied,
(ii) at least one important property is satisfied,
(iii) all satisfied properties are important,
(iv) at least one non satisfied property is not important.

Note also the cube is compatible with a bipolar understanding [18]. Suppose that among possible properties for the considered objects, some are desirable (or requested) and form a subset R and some others should be excluded (forbidden or undesirable) and form a subset E. Clearly, one should have $E \subseteq \overline{R}$. The set of properties of a given object is partitioned into the subset of satisfied properties S and the subset $\overline{S}$ of not satisfied properties. Then vertex **A** corresponds to $R \subseteq S$ and **a** to $\overline{R} \subseteq \overline{S}$. Then **a** also corresponds to $E \subseteq \overline{S}$.

2.2 A Gradual Cube of Opposition

It has been recently shown that the structure of the cube of opposition underlies many knowledge representation formalisms used in artificial intelligence, such as first order logic, modal logic, but also formal concept analysis, rough set theory, abstract argumentation, as well as quantitative uncertainty modeling frameworks such as possibility theory, or belief function theory [14, 19]. In order to accommodate quantitative frameworks, a graded extension of the cube has been defined in the following way.

Let $\alpha, \iota, \varepsilon, o$, and $\alpha', \iota', \varepsilon', o'$ be the grades in $[0, 1]$ associated to vertices **A**, **I**, **E**, **O** and **a**, **i**, **e**, **o**. Then we consider an involutive negation n, a symmetrical conjunction $*$ that respects the law of contradiction with respect to this negation, and we interpret entailment in the many-valued case by the inequality $\leq$: the conclusion is at least as true as the premise. The constraints satisfied by the cube of Fig. 2 can be generalized in the following way [14]:

(i) $\alpha = n(o)$, $\varepsilon = n(\iota)$ and $\alpha' = n(o')$ and $\varepsilon' = n(\iota')$;
(ii) $\alpha \leq \iota$, $\varepsilon \leq o$ and $\alpha' \leq \iota'$, $\varepsilon' \leq o'$;
(iii) $\alpha * \varepsilon = 0$ and $\alpha' * \varepsilon' = 0$;
(iv) $n(\iota) * n(o) = 0$ and $n(\iota') * n(o') = 0$;
(v) $\alpha \leq \iota'$, $\alpha' \leq \iota$ and $\varepsilon' \leq o$, $\varepsilon \leq o'$;
(vi) $\alpha' * \varepsilon = 0$, $\alpha * \varepsilon' = 0$;
(vii) $n(\iota') * n(o) = 0$, $n(\iota) * n(o') = 0$.

In the paper, we restrict to the numerical setting and let $n(a) = 1 - a$. It leads to define $* = \max(0, \cdot + \cdot - 1)$ (the Łukasiewicz conjunction). In the sequel, we show that the (gradual) cube of opposition is relevant for describing different families of multiple criteria aggregation functions. We first illustrate this fact by considering weighted minimum and maximum, together with related aggregations.

3 A Cube of Simple Qualitative Aggregations

In multiple criteria aggregation objects are evaluated by means of criteria i where $i \in \mathscr{C} = \{1, \ldots, n\}$. The evaluation scale L is a totally ordered scale with top 1, bottom 0, and the order-reversing operation is denoted by $1 - (\cdot)$. For simplicity, we take $L = [0, 1]$, or a subset thereof, closed under the negation and the conjunction.

An object x is represented by a vector $x = (x_1, \ldots, x_n)$ where x_i is the evaluation of x according to the criterion i. We assume that $x_i = 1$ means that the object fully satisfies criterion i and $x_i = 0$ expresses a total absence of satisfaction. Let $\pi_i \in [0, 1]$ represent the level of importance of criterion i. The larger π_i the more important the criterion. We note $\pi = (\pi_1, \ldots, \pi_n)$.

In such a context, simple qualitative aggregation operators are the weighted min and the weighted max [20]:

- The weighted min measures the extent to which all important criteria are highly satisfied; it corresponds to the expression $\min_{i=1}^{n} \max(1 - \pi_i, x_i)$,
- the weighted max, $\max_{i=1}^{n} \min(\pi_i, x_i)$, is optimistic and only requires that at least one important criterion be highly satisfied.

The weighted min and weighted max correspond to vertices **A** and **I** of the cube on Fig. 3. As it can be noticed, the cube of Fig. 3 is just a multiple-valued counterpart of the initial cube of Fig. 2.

Under the hypothesis of the double normalization ($\exists i, \pi_i = 1$ and $\exists j, \pi_j = 0$) and the hypothesis $\exists r, x_r = 1$ and $\exists s, x_s = 0$, which correspond to the non-emptiness of P, $\overline{P}$, Q, and $\overline{Q}$ in cube of Fig. 2, it can be checked that all the constraints (i–vii) of the gradual cube hold. For instance, the entailment from **A** to **I** translates into

Fig. 3 The cube of
weighted qualitative
aggregations

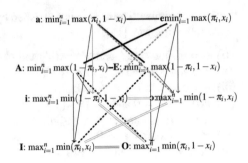

$\min_{i=1}^n \max(1 - \pi_i, x_i) \leq \max_{i=1}^n \min(\pi_i, x_i)$, which holds as soon as $\exists i, \pi_i = 1$. Formally speaking, in terms of possibility theory [21, 22], it is nothing but the expression that the necessity of a fuzzy event $N_\pi(x)$ is less or equal to the possibility $\Pi_\pi(x)$ of this event, provided that the possibility distribution π is normalized. While **A** and **I** are associated with $N_\pi(x)$ and $\Pi_\pi(x)$ respectively, **a** is associated with a guaranteed possibility $\Delta_\pi(x)$ (which indeed reduces to $\Delta_\pi(x) = \min_{i \mid x_i=1} \pi_i$ in case $\forall i, x_i \in \{0, 1\}$). Note also that $\Delta_\pi(x) = N_{\bar\pi}(1 - x)$, where $\bar\pi = 1 - \pi_i$; lastly **i** is associated with $\nabla_\pi(x) = 1 - \Delta_\pi(1 - x)$. Moreover there is a correspondence between the aggregation functions on the right facet of the cube and those on the left facet, replacing x with $1 - x$.

Let us discuss the different aggregation attitudes displayed on the cube. Suppose that a fully satisfactory object x is an object with a global rating equal to 1. Then, vertices **A, I, a** and **i** correspond respectively to 4 different cases: x is such that

(i) **A**: all properties having some importance are fully satisfied (if $\pi_i > 0$ then $x_i = 1$ for all i),
(ii) **I**: there exists at least one important property i fully satisfied ($\pi_i = 1$ and $x_i = 1$),
(iii) **a**: all somewhat satisfied properties are fully important (if $x_i > 0$ then $\pi_i = 1$ for all i),
(iv) **i**: there exists at least one unimportant property i that is not satisfied at all ($\pi_i = 0$ and $x_i = 0$).

These cases are similar to those encountered in the cube of Fig. 2.

Example 1 We consider $\mathscr{C} = \{1, 2, 3\}$ and $\pi_1 = 0, \pi_2 = 0.5$ and $\pi_3 = 1$; see Fig. 4.

- on vertex A (resp. I) a fully satisfied object is such that $x_2 = x_3 = 1$ (resp. $x_3 = 1$),
- on vertex a (resp. i) a fully satisfied object is such that $x_1 = x_2 = 0$ (resp. $x_1 = 0$).

The operations of the front facet of the cube of Fig. 3 merge positive evaluations that focus on the high satisfaction of important criteria, while the local ratings x_i on the back could be interpreted as negative ones (measuring the intensity of faults). Then, aggregations yield global ratings evaluating the lack of presence of important faults. In this case, weights are tolerance levels forbidding a fault to be too strongly present. Then, the vertices a and i in the back facet are interpreted differently:

Fig. 4 Example of a cube of
weighted qualitative
aggregations

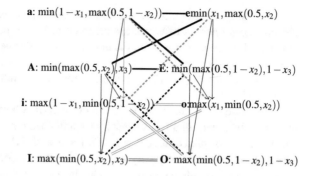

a: $\min(1-x_1, \max(0.5, 1-x_2))$ — e: $\min(x_1, \max(0.5, x_2))$

A: $\min(\max(0.5, x_2), x_3)$ — E: $\min(\max(0.5, 1-x_2), 1-x_3)$

i: $\max(1-x_1, \min(0.5, 1-x_2))$ — o: $\max(x_1, \min(0.5, x_2))$

I: $\max(\min(0.5, x_2), x_3)$ — O: $\max(\min(0.5, 1-x_2), 1-x_3)$

- the evaluation associated to a is equal to 1 if all somewhat intolerable faults are fully absent;
- the evaluation associated to i is equal to 1 if there exists at least one intolerable fault that is absent.

This framework thus involves two complementary points of view, recently discussed in a multiple criteria aggregation perspective [11].

4 The Cube of Sugeno Integrals

Weighted minimum and maximum (as well as ordered weighted minimum and maximum [23]) are particular cases of Sugeno integrals. The cube on Fig. 3 can indeed be extended to Sugeno integrals and its associated so-called desintegrals. Before presenting the cube associated with Sugeno integrals, let us recall some definitions used in the following, namely the notions of capacity, conjugate capacity, qualitative Moebius transform, and focal sets.

In the definition of a Sugeno integral the relative weights of the set of criteria are represented by a capacity (or fuzzy measure) which is a set function $\mu : 2^{\mathscr{C}} \to L$ that satisfies $\mu(\emptyset) = 0$, $\mu(\mathscr{C}) = 1$ and $A \subseteq B$ implies $\mu(A) \leq \mu(B)$. The conjugate capacity of μ is defined by $\mu^c(A) = 1 - \mu(\overline{A})$ where $\overline{A}$ is the complement of A.

The inner qualitative Moebius transform of a capacity μ is a mapping $\mu_\# : 2^{\mathscr{C}} \to L$ defined by

$$\mu_\#(E) = \mu(E) \text{ if } \mu(E) > \max_{B \subset E} \mu(B) \text{ and } 0 \text{ otherwise.}$$

A set E such that $\mu_\#(E) > 0$ is called a focal set. The set of the focal sets of μ is denoted by $\mathscr{F}(\mu)$.

The Sugeno integral of an object x with respect to a capacity μ is originally defined by [24, 25]:

$$S_\mu(x) = \max_{\alpha \in L} \min(\alpha, \mu(\{i \mid x_i \geq \alpha\})). \tag{1}$$

When Sugeno integrals are used as aggregation functions to select acceptable objects, the properties of which are assumed to have a positive flavor: namely, the global evaluation increases with the partial ratings. But generally, we may also have negative properties, as already described in the introduction. In such a context we can use a desintegral [11, 12] associated to the Sugeno integral. We now present this desintegral.

In the case of negative properties, fault-tolerance levels are assigned to sets of properties by means of an anti-capacity (or anti-fuzzy measure), which is a set function $v : 2^{\mathscr{C}} \to L$ such that $v(\emptyset) = 1$, $v(\mathscr{C}) = 0$, and if $A \subseteq B$ then $v(B) \leq v(A)$. The conjugate v^c of an anti-capacity v is an anti-capacity defined by $v^c(A) = 1 - v(\overline{A})$, where $\overline{A}$ is the complementary of A. The desintegral $S_v^{\downarrow}(x)$ is defined from the corresponding Sugeno integral, by reversing the direction of the local value scales (x becomes $1 - x$), and by considering a capacity induced by the anti-capacity v, as follows:

$$S_v^{\downarrow}(x) = S_{1-v^c}(1 - x). \tag{2}$$

In order to present the square of Sugeno integrals, we need to define the pessimistic part and the optimistic part of a capacity. They are respectively called assurance and opportunity functions by Yager [26]. This need should not come as a surprise. Indeed the entailment from **A** to **I** requires that the expression in **A** have a universal flavor, i.e. here, is minimum-like, while the expression in **I** have an existential flavor, i.e. here, is maximum-like, but the capacity μ, on which the considered Sugeno integral is based, may have neither.

When we consider a capacity μ, its pessimistic part is $\mu_*(A) = \min(\mu(A), \mu^c(A))$ and its optimistic part is $\mu^*(A) = \max(\mu(A), \mu^c(A))$ [13]. Observe that $\mu_* \leq \mu^*$, $\mu_*{}^c = \mu^*$ and $\mu^{*c} = \mu_*$. So a capacity μ induces the following square of opposition (see [27] for more details).

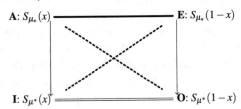

Note that $S_{\mu_*}(1 - x) = S_{1-\mu^*}^{\downarrow}(x)$ and $S_{\mu^*}(1 - x) = S_{1-\mu_*}^{\downarrow}(x)$, where $1 - \mu^*$ and $1 - \mu_*$ are anti-capacities.

Lastly, in order to build the cube associated to Sugeno integrals, just as $\overline{\pi}$ is at work on the back facet of the cube associated with weighted min and max, we also need the opposite capacity $\overline{\mu}$, defined as follows: $\overline{\mu_\#}(E) = \mu_\#(\overline{E})$ and $\overline{\mu}(A) = \max_{\overline{E} \subseteq A} \mu_\#(E)$. A square of opposition **aieo** can be defined with the capacity $\overline{\mu}$. Hence, supposing $\exists i \in \mathscr{C}$ such that $x_i = 0$ and $\exists j \in \mathscr{C}$ such that $x_j = 1$, we can construct a cube of opposition **AIEO** and **aieo** as presented in Fig. 5 [14].

Fig. 5 Cube of opposition
of Sugeno integrals
associated to a capacity μ

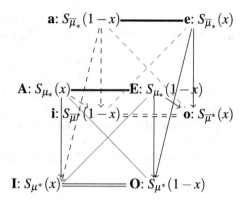

The fact that all the constraints of a gradual cube hold in this case has been only established under a specific type of normalization for capacities [27], i.e., $\exists A \neq \mathscr{C}$ such that $\mu(A) = 1$ and $\exists B \neq \mathscr{C}$ such that $\mu^c(B) = 1$; note that in such a context there exists a non empty set, $\overline{B}$, such that $\mu(\overline{B}) = 0$. However, this does not cover another particular case where the constraints of the cube also hold, namely the one where μ is only non zero on singletons. Finding the most general condition on μ ensuring the satisfaction of all constraints (i–vii) in the cube of Sugeno integrals is still an open question.

Let us now present the aggregation attitudes expressed by the cube of Sugeno integrals. We can characterize situations where objects get a global evaluation equal to 1 using aggregations on the side facet.

The global evaluations at vertices **AIai** of a cube associated to a capacity μ are maximal respectively in the following situations pertaining to the focal sets of μ:

A The set of totally satisfied properties contain a focal set with weight 1 and over-laps all other focal sets.

I The set of satisfied properties contains a focal set with weight 1 or overlaps all other focal sets.

a The set of totally violated properties contains no focal set and its complement is contained in a focal set with weight 1.

i The set of totally violated properties contains no focal set or its complement is contained in a focal set with weight 1.

Example 2 Assume $\mathscr{C} = \{1, 2, 3\}$ and the following capacities

Capacity	{1}	{2}	{3}	{1,2}	{1,3}	{2,3}	{1,2,3}
μ	0	0	0	1	1	0	1
μ^c	1	0	0	1	1	1	1
$\overline{\mu}$	0	1	1	1	1	1	1
$\overline{\mu}^c$	0	0	0	0	0	1	1

$\mu^c \geq \mu$ so $\mu_* = \mu$ and $\mu^* = \mu^c$
$\overline{\mu} \geq \overline{\mu}^c$ so $\overline{\mu}^* = \overline{\mu}$ and $\overline{\mu}_* = \overline{\mu}^c$
Note that $\overline{\mu}$ is a possibility measure.

The aggregation functions on the vertices are:

A: $S_\mu(x) = \max(\min(x_1, x_2), \min(x_1, x_3))$, **I**: $S_{\mu^c}(x) = \max(x_1, \min(x_2, x_3))$

a: $S_{\overline{\mu^c}}(1-x) = \min(1-x_2, 1-x_3)$, **i**: $S_{\overline{\mu}}(1-x) = \max(1-x_2, 1-x_3)$.

- For vertex **A**, the two focal sets overlap when $S_\mu(x) = 1$.
- For vertex **I**, one can see that $S_{\mu^c}(x) = 1$ when $x_1 = 1$ and $\{1\}$ does overlap all focal sets of μ; the same occurs when $x_2 = x_3 = 1$.
- For vertex **a**, $S_{\overline{\mu^c}}(1-x) = 1$ when $x_2 = x_3 = 0$, and note that the complement of $\{2, 3\}$ is contained in a focal set of μ, while $\{2, 3\}$ contains no focal set of μ.
- For vertex **i**, $S_{\overline{\mu}}(1-x) = 1$ when, $x_2 = 0$ or $x_3 = 0$, and clearly, neither $\{2\}$ not $\{3\}$ contain any focal set of μ, but the complement of each of them is a focal set of μ.

5   The Cube of Choquet Integrals

When criteria evaluations are quantitative, Choquet integrals often constitute a suitable family of aggregation operators, which generalize weighted averages, and which parallel, in different respects, the role of Sugeno integrals for the qualitative case. Although the evaluation scale can be taken as the real line $\mathbb{R}$, we use the unit interval $[0, 1]$ in the following.

Belief and plausibility functions are particular cases of Choquet integrals, just as necessity and possibility measures are particular cases of Sugeno integrals. This is why we begin with the presentation of the cube of belief functions, before studying the cube of Choquet integrals, of which another noticeable particular case is the cube of ordered weighted averaging aggregation operators (*OWA*), which is then discussed, before concluding.

5.1   The Cube of Belief Functions

In Shafer's evidence theory [28], a belief function Bel_m is defined together with a dual plausibility function Pl_m from a mass function m, i.e., a real set function $m : 2^{\mathscr{C}} \to [0, 1]$ such that $m(\emptyset) = 0$, $\sum_{A \subseteq \mathscr{C}} m(A) = 1$. Then for $A \subseteq \mathscr{C}$, we have $Bel_m(A) = \sum_{E \subseteq A} m(E)$ and $Pl_m(A) = 1 - Bel_m(\overline{A}) = \sum_{E \cap A \neq \emptyset} m(E)$.

Viewing m as a random set, the complement $\overline{m}$ of the mass function m is defined as $\overline{m}(E) = m(\overline{E})$ [29]. The commonality function Q and its dual $\eth$ are then defined by $Q_m(A) = \sum_{A \subseteq E} m(E)$ and $\eth_m(A) = \sum_{\overline{E} \cap \overline{A} \neq \emptyset} m(E) = 1 - Q_m(\overline{A})$ respectively. The normalization $\overline{m}(\emptyset) = 0$ forces $m(\mathscr{C}) = 0$. Then, $Q_m(A) = Bel_{\overline{m}}(\overline{A})$ while $\eth_m(A) = Pl_{\overline{m}}(\overline{A})$. It can be checked that the transformation $m \to \overline{m}$ reduces to $\pi \to \overline{\pi} = 1 - \pi$ in case of nested focal elements. All these set functions can be put on the following cube of opposition [14]. See Fig. 6. Indeed, if $m(\emptyset) = 0$, we have $Bel_m(A) \leq Pl_m(A) \Leftrightarrow Bel_m(A) + Bel_m(\overline{A}) \leq 1 \Leftrightarrow Pl_m(A) + Pl_m(\overline{A}) \geq 1$, which gives birth to the square of

Fig. 6 Cube of opposition
of evidence theory

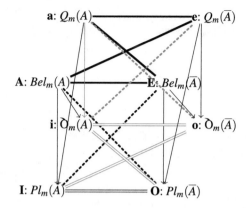

opposition **AIEO**. We can check as well that $Bel_m(A) = \sum_{E \subseteq A} m(E) \leq \eth_m(A) = 1 - \sum_{\overline{A} \subseteq E} m(E)$. Similar inequalities ensure that $Q_m(A) \leq Pl_m(A) = 1 - Bel_m(\overline{A})$, or $Bel_m(A) + Q_m(\overline{A}) \leq 1$, for instance, which ensures that the constraints of the cube hold.

Belief functions are a particular case of capacities. Note that the square can be extended replacing *Bel* and *Pl* by a capacity μ and its conjugate $\mu^c(A) = 1 - \mu(\overline{A})$, respectively. However, to build the cube, we also need inequalities such as $Q_m(A) \leq Pl_m(A)$ to be generalized to capacities.

5.2 Extension to Choquet Integrals

Considering a capacity μ on $\mathscr{C}$, the Moebius transform of μ, denoted by m_μ, is given by $m_\mu(T) = \sum_{K \subseteq T}(-1)^{|T \setminus K|}\mu(K)$. The Choquet integral with respect to μ is:

$$C_\mu(x) = \sum_{T \subseteq \mathscr{C}} m_\mu(T) \min_{i \in T} x_i. \tag{3}$$

Clearly, $Bel_m(A) = \sum_E m(E) \cdot \min_{u \in E} 1_A(u)$. We have the equality $Bel_{m_\mu}(A) = C_\mu(1_A) = \mu(A)$ if m_μ represents the Moebius transform of a capacity μ. This characterisation is presented in [30]. More precisely, a real set function m is the Moebius transform of a capacity μ if and only if $m(\emptyset) = 0$, $\sum_{K \subseteq S} m(i \cup K) \geq 0$ for all i and for all $S \subseteq \mathscr{C} \setminus i$ and $\sum_{K \subseteq \mathscr{C}} m(K) = 1$. And it is the Moebius transform of a belief function if and only if it is non-negative. In general, $m_\mu(E)$ can be negative for non-singleton sets.

Under these conditions, $Pl_m(A) = 1 - C_\mu(1_{\overline{A}})$. But we have $Q_m(A) = Bel_{\overline{m}}(\overline{A})$, so $Q_m(A) = C_{\overline{\mu}}(1_{\overline{A}})$ if $\overline{m}$ satisfies the conditions to be a Moebius transform of a capacity $\overline{\mu}$. In such a context, $\eth_m(A) = Pl_{\overline{m}}(\overline{A}) = 1 - C_{\overline{\mu}}(1_A)$. It is worth noticing that there

Fig. 7 Cube of opposition
of Choquet integral

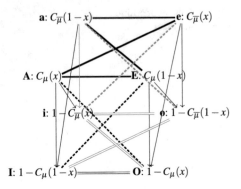

exist Moebius transforms m such that $\overline{m}$ is not a Moebius transform since we need to have the condition $m(\mathscr{C}) = 0$.

Hence one may consider the extension of the cube of Fig. 6 to general Choquet integrals. In order to understand the proof of the following proposition we need the other expression of the Choquet integral: $C_\mu(x) = \sum_{i=1}^{n}(x_i - x_{i-1})\mu(A_i)$ where we suppose that $x_1 \leq \cdots \leq x_n$, $A_i = \{i, \ldots, n\}$ and $x_0 = 0$.

With this expression, it is easy to check that the Choquet integral is increasing according to the capacity. Then the following holds (See Fig. 7).

Proposition 1 *The cube of Choquet integral is a cube of opposition if and only if* $\mu \leq \mu^c$, $\overline{\mu} \leq \overline{\mu}^c$ *and* $\mu + \overline{\mu} \leq 1$.

Proof We consider the evaluation scale $[0, 1]$. Without loss generality we can suppose that $x_1 \leq \cdots \leq x_n$.

A entails **I** iff $C_\mu(x) + C_\mu(1 - x) \leq 1$. Considering $x = 1_A$ the characteristic function of A we need $\mu(A) \leq \mu^c(A)$. If $\mu \leq \mu^c$ then $C_\mu(x) \leq C_{\mu^c}(x)$. We have $C_\mu(1 - x) = \sum_{T \subseteq \mathscr{C}} m_\mu(T) \min_{i \in T}(1 - x_i) = \sum_{T \subseteq \mathscr{C}}(m_\mu(T) - m_\mu(T) \max_{i \in T} x_i) = 1 - C_{\mu^c}(x)$. So, $C_\mu(x) + C_\mu(1 - x) \leq C_\mu(x) + C_{\mu^c}(1 - x) = 1$.

By symmetry we have **E** entails **O**.

A and **E** cannot be equal to 1 together: $C_\mu(x) = 1$ entails $C_{\mu^c}(x) = 1$ since $\mu \leq \mu^c$, i.e., $C_\mu(1 - x) = 0$. By duality **I** and **O** cannot be equal to 0 together.

So **AEIO** is a square of opposition.

Similarly $\overline{\mu} \leq \overline{\mu}^c$ is equivalent to making **aeio** a square of opposition.

If $C_\mu(x) \leq 1 - C_{\overline{\mu}}(x)$ then considering $x = 1_A$ we have $\mu(A) + \overline{\mu}(A) \leq 1$. Conversely if we suppose that $\mu + \overline{\mu} \leq 1$ then $C_\mu(x) + C_{\overline{\mu}}(x) = \sum_{i=1}^{n}(x_i - x_{i-1})(\mu(A_i) + \overline{\mu}(A_i)) \leq \sum_{i=1}^{n}(x_i - x_{i-1}) = x_n \leq 1$. This last equivalence permits to conclude that the considered cube is a cube of opposition.

The condition $\mu + \overline{\mu} \leq 1$ is valid for belief functions since $\overline{Bel}_m(A) = Bel_{\overline{m}}(A) = Q_m(\overline{A})$, and $Bel_m(A) + Q_m(\overline{A}) \leq 1$, but it needs to be investigated for more general capacities since some masses may be negative. Note that, in its back facet, the cube of Choquet integrals exhibits what maybe called desintegrals, associated to Choquet integrals. Namely, using $C_{\overline{\mu}}(1-x)$, the global evaluation increases when partial ratings decrease.

Let us discuss the aggregation attitudes when the evaluation scale is the real interval $[0, 1]$ and μ is a belief function. More precisely we are going to characterize the situations where an object x gets a perfect global evaluation, i.e., a global evaluation equal to 1, for the different vertices **AIai**. We denote $\mathscr{F}_\mu$ the family of the sets having a Moebius transform not equal to 0.

- **A**: $C_\mu(x) = 1$ can be written $\sum_{F \subseteq \mathscr{C}} m_\mu(F) \cdot \min_{i \in F} x_i = 1$, which implies that $\forall F \in \mathscr{F}_\mu, \forall i \in F, x_i = 1$. So the focal sets of μ are included in the set of totally satisfied properties.

- **I**: $1 - C_\mu(1-x) = 1 = C_{\mu^c}(x)$ is equivalent to $\sum_{F \subseteq \mathscr{C}} m_\mu(F) \cdot \max_{i \in F} x_i = 1$. So in this case $\forall F \in \mathscr{F}_\mu, \exists i \in F$ such that $x_i = 1$. So each focal set of μ must intersect the set of totally satisfied properties.

- **a**: $C_{\overline{\mu}}(1-x) = \sum_{F \subseteq \mathscr{C}} \overline{m_\mu}(F) \cdot \min_{i \in F}(1-x_i) = \sum_{F \subseteq \mathscr{C}} m_\mu(F) \cdot \min_{i \in \overline{F}}(1-x_i)$. Then $C_{\overline{\mu}}(1-x) = 1$ is equivalent to $\forall F \in \mathscr{F}_\mu, \min_{i \in \overline{F}}(1-x_i) = 1$, or equivalently, $\forall F \in \mathscr{F}_\mu, \max_{i \in \overline{F}} x_i = 0$, which means $\forall F \in \mathscr{F}_\mu, \forall i \notin F, x_i = 0$.
  So all properties outside each focal set of μ are violated. The only properties that are satisfied are those in the intersection of the focal sets of μ.

- **i**: we have $1 - C_{\overline{\mu}}(x) = \sum_{F \subseteq \mathscr{C}} m_\mu(F) \cdot \max_{i \in \overline{F}}(1-x_i)$. Then $1 - C_{\overline{\mu}}(x) = 1$ is equivalent to $\forall F \in \mathscr{F}_\mu, \max_{i \in \overline{F}}(1-x_i) = 1$, i.e., $\forall F \in \mathscr{F}_\mu, \min_{i \in \overline{F}} x_i = 0$, which means $\forall F \in \mathscr{F}_\mu, \exists i \notin F$ such that $x_i = 0$. So there must be at least one violated property outside each focal set of μ.

5.3 Example for the Cube of Choquet Integral

Let us consider the menu of a traditional restaurant in Lyon.[1] We leave it in French (due to the lack of precise equivalent terms in English for most dishes):

Starter

Saladier lyonnais: museau, pieds de veau, cervelas, lentilles, pommes de terre, saucisson pistaché, frisée, oreilles de cochon

---

[1] This example is specially dedicated to Ron Yager in remembrance of a dinner in Lyon in a traditional restaurant, which took place at the occasion of the CNRS Round Table on Fuzzy Sets organized by Robert Féron [31] in Lyon on June 23–25, 1980 [32]. This Round Table was an important meeting for the development of fuzzy set research, because most of the active researchers of the field were there. Interestingly enough, Robert Féron had the remarkable intuition to invite Gustave Choquet in the steering committee, at a time where no fuzzy set researcher was mentioning Choquet integrals! This meeting also included, as usual, some nice moments of relaxation and good humor. In particular, at the above-mentioned dinner, to which quite a number of people took part (including two of the authors of this paper), Ron enjoyed very much a pigs feet dish. He was visibly very happy with his choice, so Lotfi Zadeh told him, "Ron, you should have been a pig in another life", to which Ron replied "no, Lotfi, it is in this life", while continuing to suck pigs' bones with the greatest pleasure.

Oeuf meurette: oeuf poché, crotons, champignons, sauce vin rouge et lardons
Harengs pommes de terre à l'huile

Main course

Gratin d'andouillettes, sauce moutarde
Rognons de veau au Porto et moutarde
Quenelles de brochet, sauce Nantua et riz pilaf

Dessert

Gnafron: sorbet cassis et marc de Bourgogne
Baba au rhum et chantilly
Crème caramel

A tourist wants to eat some typical dishes of Lyon. His preferred dishes are "saladier lyonnais" (which offers a great sampling of meats from the Lyon region) and "gratin d'andouillettes" since he wants to eat some gourmet delicatessen products. The evaluation scale is the real interval $[0, 1]$, so the "saladier lyonnais" and "gratin d'andouillettes" get the maximal rating 1. The other dishes receive a smaller rating. The set of criteria is $\mathscr{C} = \{s, c, d\}$, where s, c, d refer to starter, main course, and dessert respectively. We consider the Möbius transform: $m : 2^{\mathscr{C}} \rightarrow [0, 1]$ defined by $m(s) = m(s, c) = 0.5$ and 0 otherwise. Such a weighting clearly stresses the importance of the starter, and acknowledges the fact that the main course is only of interest with a starter, while dessert is not an issue for this tourist. A chosen menu is represented by a vector (x_s, x_c, x_d) where x_i is the rating corresponding to the chosen dish for the criterion i. The Choquet integral of x with respect to the capacity μ associated to m is:

$$C_\mu(x) = 0.5x_s + 0.5 \cdot \min(x_s, x_c).$$

$\overline{m}$ is the set function defined by $\overline{m}(d) = \overline{m}(c, d) = 0.5$ and 0 otherwise. It is easy to check that $\overline{m}$ is a Möbius transform. The Choquet integral of x with respect to $\overline{\mu}$, the capacity defined with $\overline{m}$ is:

$$C_{\overline{\mu}}(x) = 0.5x_d + 0.5 \cdot \min(x_c, x_d).$$

Let us look at the choices that get a perfect global evaluation on the cube of Choquet integrals:

- **A**: $C_\mu(x) = 1$ iff $x_s = x_c = 1$: a menu with a maximal evaluation contains the "saladier lyonnais" and the "gratin d'andouillette."
- **I**: $C_\mu(1 - x) = 0$ iff $x_s = 1$ or $x_s = x_c = 1$: a menu with a maximal evaluation contains the "saladier lyonnais" and may contain the "gratin d'andouillette."
- **a**: $C_{\overline{\mu}}(1 - x) = 1$ iff $x_c = x_d = 0$: a menu with a maximal evaluation contains neither the "gratin d'andouillette", nor the best dessert.
- **i**: $C_{\overline{\mu}}(x) = 0$ iff $x_d = 0$ ou $x_c = x_d = 0$: a menu with a maximal evaluation does not contain the best dessert, but may contain the "gratin d'andouillette".

Without surprise, the Choquet integral in **A** is maximal if the menu includes both the "saladier lyonnais" and the "gratin d'andouillette," while **I** is maximal as soon as the menu includes at least the "saladier lyonnais". The maximality conditions in **a** (and in **i**) are less straightforward to understand. Here we should remember that already in cube of Fig. 2, **a** entails **I** provided that x is normalized (i.e., $\exists i, x_i = 1$), which ensures that the expression attached to **a** is smaller or equal to the one associated with **I**. The same condition is enough for having

$$C_{\bar{\mu}}(1-x) = \sum_{F \subseteq \mathscr{C}} m_{\mu}(F) \cdot \min_{i \in \bar{F}} 1 - x_i \leq \quad 1 - \quad C_{\mu}(1-x) = \sum_{F \subseteq \mathscr{C}} m_{\mu}(F) \cdot$$

$\max_{i \in F} x_i$ provided that $\sum_{F \subseteq \mathscr{C}} m_{\mu}(F) = 1$. Indeed, let $x_{i^*} = 1$, then for all $F \subseteq \mathscr{C}$, either $x_{i^*} \in F$ or $x_{i^*} \in \bar{F}$. Thus, either $\min_{i \in \bar{F}} 1 - x_i = 0$, or $\max_{i \in F} x_i = 1$, which ensures the inequality.

Thus going back to the example, since the evaluation in **a** is maximal for $x_c = x_d = 0$, the normalization forces $x_s = 1$, which means that the menu includes the "saladier lyonnais". Note also that $x_s = 1, x_c = 0, x_d = 0$ is a minimal normalized evaluation vector x, for which the desintegral associated with **a** is maximal. Considering the evaluation in **i** the normalization entails that $x_s = 1$ or $x_c = 1$ so the menu includes the "saladier lyonnais" or the "gratin d'andouillette".

5.4 The Cube of OWA Operators

Ordered Weighted Averages (OWA) [8–10] and their weighted extension [33] have been found useful in many applications. Since OWAs are a particular case of Choquet integrals [34], one may wonder about a square, and then a cube of opposition associated to OWAs as a particular case of the cube of Fig. 7. Let us first recall what is an OWA.

An OWA_w is a real mapping on $\mathscr{C}$ associated to a collection of weights $w = (w_1, \ldots, w_n)$ such that $w_i \in [0, 1]$ for all $i \in \{1, \ldots, n\}$, $\sum_{i=1}^{n} w_i = 1$, and defined by:

$$OWA_w(x) = \sum_{i=1}^{n} w_i \cdot x_{(i)}$$

where $x_{(1)} \leq \cdots \leq x_{(n)}$.

This includes noticeable particular cases:

- $w = (1, 0, \ldots, 0) \Rightarrow OWA_w(x) = \min_{i=1}^{n} x_i$,
- $w = (0, \ldots, 0, 1) \Rightarrow OWA_w(x) = \max_{i=1}^{n} x_i$,
- $w = (\frac{1}{n}, \ldots, \frac{1}{n}) \Rightarrow OWA_w(x) = \frac{\sum_{i=1}^{n} x_i}{n}$.

In [8], Yager also defines measures of orness and andness:

$$orness(OWA_w) = \frac{1}{n-1} \sum_{i=1}^{n} (n-i) \cdot w_i; \quad andness(OWA_w) = 1 - orness(OWA_w).$$

Fig. 8 Square of opposition
of OWA

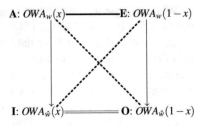

Note that $orness(OWA_w), andness(OWA_w) \in [0, 1]$. The closer the OWA_w is to an *or* (resp. *and*), the closer $orness(OWA_w)$ is to 1 (resp. 0).

In the same article, Yager also defines the measure of dispersion (or entropy) of an OWA associated to w by

$$disp(OWA_w) = - \sum_{i=1}^{n} w_i \ln w_i.$$

The measure of dispersion estimates the degree to which we use all the aggregates equally.

The dual of OWA_w (see, e.g., [35]) is $OWA_{\hat{w}}$ with the weight $\hat{w} = (w_n, \dots, w_1)$. More precisely we have $\hat{w}_i = w_{n-i+1}$. It is easy check that $disp(OWA_{\hat{w}}) = disp(OWA_w)$ and $orness(OWA_{\hat{w}}) = 1 - orness(OWA_w) = andness(OWA_w)$.

The following duality relation holds

$$OWA_w(1 - x) = \sum_{i=1}^{n} w_i(1 - x_{(n-i+1)}) = 1 - \sum_{i=1}^{n} w_{n-i+1}x_{(i)}$$
$$= 1 - OWA_{\hat{w}}(x)$$

In particular, it changes min into max and conversely.

This corresponds to the expected relation for the diagonals of the square of opposition of Fig. 8 for OWAs. Then the entailment relations of the vertical sides require to have

$$\sum_{i=1}^{n} w_i \cdot x_{(i)} \leq \sum_{i=1}^{n} w_i \cdot x_{(n-i+1)}$$

This can be rewritten as

$$0 \leq w_1 \cdot (x_{(n)} - x_{(1)}) + w_2 \cdot (x_{(n-1)} - x_{(2)}) + \cdots + w_n \cdot (x_{(1)} - x_{(n)})$$
$$= (w_1 - w_n) \cdot (x_{(n)} - x_{(1)}) + (w_2 - w_{n-1}) \cdot (x_{(n-1)} - x_{(2)}) + \cdots$$

In order to guarantee that the above sum adds positive terms only, it is enough to enforce the following condition for the weights:

$$w_1 \geq w_2 \geq \cdots \geq w_n,$$

which expresses a demanding aggregation. We are not surprised to observe that the w associated to max violates the above condition. The situation is similar to the one already encountered with Sugeno integrals where we had to display integrals based on pessimistic or optimistic fuzzy measures depending on the vertices of the square and similar to the situation of belief functions, which are pessimistic, which ensures a regular square of opposition without any further condition.

Besides, in [34, 36–38] it is proved that a capacity μ depends only on the cardinality of subsets if and only if there exists $w \in [0, 1]^n$ such that $C_\mu(x) = OWA_w(x)$. Moreover we have the following relations. The fuzzy measure μ associated to OWA_w is given by: $\mu(T) = \sum_{i=n-t+1}^{n} w_i$ where t denotes the cardinality of T. It is worth noticing that the Moebius transform is $m(T) = \sum_{j=0}^{t-1} \binom{t-1}{j} (-1)^{t-1-j} w_{n-j}$, so m depends only on the cardinality of the subsets. It is worth noticing that while the particular cases min and average are associated with simple positive mass functions ($m(\mathscr{C}) = 1$, and $m(\{i\}) = 1/n$ respectively), max is associated with a mass function that has negative weights (remember that plausibility measures do not have a positive Moebius transform).

Conversely we have $w_{n-t} = \mu(T \cup i) - \mu(T) = \sum_{K \subseteq T} m(K \cup i)$ $i \in \mathscr{C}$ $T \subseteq \mathscr{C} \backslash i$. So if μ depends only on the cardinality of the subsets, $\overline{\mu}$, the capacity associated to $\overline{m}$, depends only on the cardinality of subsets (since the Moebius transform depends only on the cardinality of subsets). The weight of the OWA associated to $\overline{\mu}$: $\overline{w}_{n-t} = \overline{\mu}(T \cup i) - \overline{\mu}(T)$. Moreover, note that $m(T)$ involves weights from w_{n-t+1} to w_n, while $\overline{m}(T) = \sum_{j=0}^{n-t-1} \binom{n-t-1}{j} (-1)^{n-t-1-j} w_{n-j}$ involves weights from w_{t+1} to w_n, and $\hat{m}(T) = \sum_{j=0}^{t-1} \binom{t-1}{j} (-1)^{t-1-j} \hat{w}_{n-j}$ involves weights from w_0 to w_t, since $\hat{w}_{n-j} = w_{j+1}$. This indicates that these mass functions are different.

Hence we obtain the cube associated to the OWA's presented on Fig. 9.

A deeper investigation of this cube in relation with conditions ensuring entailments from top facet to bottom facet, and the positivity of associated mass functions is left for further research.

Fig. 9 The cube of opposition for OWA operators

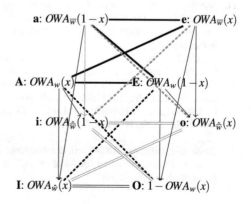

6 Concluding Remarks

This paper has first shown how the structure of the cube of opposition extends from ordinary sets to weighted min- and max-based aggregations and more generally to Sugeno integrals, which constitute a very important family of qualitative aggregation operators. Then, a similar construct has been exhibited for Choquet integrals and OWA operators. The cube exhausts all the possible aggregation attitudes. Moreover, as mentioned in Sect. 2, it is compatible with a bipolar view where we distinguish between desirable properties and rejected properties. It thus provides a rich theoretical basis for multiple criteria aggregation. Still further research is needed for a better understanding of the interplay of the vertices in the different cubes.

References

1. Zadeh, L.A.: Fuzzy sets. Inf. Control **8**, 338–353 (1965)
2. Bellman, R.E., Zadeh, L.A.: Decision-making in a fuzzy environment. Manage. Sci. **17**(4), B-141–B-164 (1970)
3. Klement, E.P., Mesiar, R., Pap, E.: Triangular Norms. Kluwer, Dordrecht (2000); Springer, Berlin (2013)
4. de Campos, L.M., Bolaños, M.J.: Characterization and comparison of Sugeno and Choquet integrals. Fuzzy Sets Syst. **52**, 61–67 (1992)
5. Grabisch, M., Labreuche, Ch.: A decade of application of the Choquet and Sugeno integrals in multi-criteria decision aid. 4OR **6**, pp. 1–44 (2008)
6. Yager, R.R.: On a general class of fuzzy connectives. Fuzzy Sets Syst. **4**(3), 235–242 (1980)
7. Yager, R.R., Rybalov, A.: Uninorm aggregation operators. Fuzzy Sets Syst. **80**(1), 111–120 (1996)
8. Yager. R.R.: On ordered weighted averaging aggregation operators in multicriteria decision making. IEEE Trans. Syst. Man Cybern. **18**(1), 183–190 (1988)
9. Yager, R.R., Kacprzyk, J. (eds.): The Ordered Weighted Averaging Operators: Theory and Applications. Kluwer Acad. Publ. (1997)
10. Yager, R.R., Kacprzyk, J., Beliakov, G. (eds.): Recent Developments in the Ordered Weighted Averaging Operators: Theory and Practice. Studies in Fuzziness and Soft Computing 265, Springer (2011)
11. Dubois, D., Prade, H., Rico, A.: Qualitative integrals and desintegrals: how to handle positive and negative scales in evaluation. In: Proceedings of 14th IPMU (Information Processing and Management of Uncertainty in Knowledge-Based Systems), vol. 299 of CCIS, pp. 306–316. Springer (2012)
12. Dubois, D., Prade, H., Rico, A.: Residuated variants of Sugeno integrals: towards new weighting schemes for qualitative aggregation methods. Inf. Sci. to appear
13. Dubois, D., Prade, H., Rico, A.: On the informational comparison of qualitative fuzzy measures. In: Proceedings of IPMU'14 (Information Processing and Management of Uncertainty in Knowledge-Based Systems), vol. 442 of CCIS, pp. 216–225 (2014)
14. Dubois, D., Prade, H., Rico, A.: The cube of opposition. A structure underlying many knowledge representation formalisms. In: Proceedings of 24th International Joint Conference on Artificial Intelligence (IJCAI'15), Buenos Aires, Jul. 25–31, pp. 2933–2939 (2015)
15. Parsons, T.: The traditional square of opposition. In: Zalta, E.N. (ed.) The Stanford Encyclopedia of Philosophy (2008)
16. Reichenbach, H.: The syllogism revised. Philos. Sci. **19**(1), 1–16 (1952)

17. Dubois, D., Prade, H.: From Blanché's hexagonal organization of concepts to formal concept analysis and possibility theory. Logica Univers. **6**, 149–169 (2012)
18. Dubois, D., Prade, H.: An introduction to bipolar representations of information and preference. Int. J. Intell. Syst. **23**(8), 866–877 (2008)
19. Ciucci, D., Dubois, D., Prade, H.: The structure of oppositions in rough set theory and formal concept snalysis—toward a new bridge between the two settings. In: Beierle, Ch., Meghini, C. (eds.) Proceedings of 8th International Symposium on Foundations of Information and Knowledge Systems (FoIKS'14), Bordeaux, Mar. 3–7, LNCS 8367, pp. 154–173. Springer (2014)
20. Dubois, D., Prade, H.: Weighted minimum and maximum operations. An addendum to 'A review of fuzzy set aggregation connectives'. Inf. Sci. **39**, 205–210 (1986)
21. Dubois, D., Prade, H.: Possibility theory and its applications: where do we stand? In: Kacprzyk, J., Pedrycz, W. (eds.) Springer Handbook of Computational Intelligence, pp. 31–60. Springer (2015)
22. Zadeh, L.A.: Fuzzy sets as a basis for a theory of possibility. Fuzzy Sets Syst. **1**(1), 3–28 (1978)
23. Dubois, D., Prade, H.: Semantics of quotient operators in fuzzy relational databases. Fuzzy Sets Syst. **78**, 89–93 (1996)
24. Sugeno, M.: Theory of fuzzy integrals and its applications. Ph.D. Thesis, Tokyo Institute of Technology, Tokyo (1974)
25. Sugeno, M.: Fuzzy measures and fuzzy integrals: a survey. In: Gupta, M.M., Saridis, G.N., Gaines, B.R. (eds.) Fuzzy Automata and Decision Processes, pp. 89–102. North-Holland (1977)
26. Yager, R.R.: Measures of assurance and opportunity in modeling uncertain information. Int. J. Intell. Syst. **27**(8), 776–797 (2012)
27. Dubois, D., Prade, H., Rico, A.: The cube of opposition and the complete appraisal of situations by means of Sugeno integrals. In: Proceedings of International Symposium on Methodologies for Intelligent Systems (ISMIS'15), Lyon, oct. 21–23, to appear (2015)
28. Shafer, G.: A Mathematical Theory of Evidence. Princeton University (1976)
29. Dubois, D., Prade, H.: A set-theoretic view of belief functions. Logical operations and approximation by fuzzy sets. Int. J. Gen. Syst. **12**(3), 193–226, (1986). Reprinted in Classic Works of the Dempster-Shafer Theory of Belief Functions. In: Yager, R.R., Liu, L. (eds.) Studies in Fuzziness and Soft Computing, vol. 219, pp. 375–410 (2008)
30. Chateauneuf, A., Jaffray, J.-Y.: Some characterizations of lower probabilities and other monotone capacities through the use of Möbius inversion. Math. Soc. Sci. **17**(3), 263–283 (1989). Reprinted in Classic Works of the Dempster-Shafer Theory of Belief Functions. In: Yager, R.R., Liu, L. (eds.) Studies in Fuzziness and Soft Computing, vol. 219, pp. 477–498 (2008)
31. Auray, J.-P., Prade, H.: Robert Féron: a pioneer in soft methods for probability and statistics. In: Proceedings of 4th International Workshop in Soft Methods in Probability and Statistics (SMPS), Toulouse, Sept. 8–10, pp. 27–32 (2008)
32. Table Ronde: Quelques Applications Concrètes Utilisant les Derniers Perfectionnements de la Théorie du Flou. BUSEFAL (LSI,Université Paul Sabatier, Toulouse), n°2, 103–105 (1980)
33. Torra, V.: The weighted OWA operator. Int. J. Intell. Syst. **12**, 153–166 (1997)
34. Fodor, J., Marichal, J.-L., Roubens, M.: Characterization of the ordered weighted averaging operators. IEEE Trans. Fuzzy Syst. **3**(2), 236–240 (1995)
35. Fodor, J., Yager, R.R.: Fuzzy set-theoretic operators and quantifiers. In: Dubois, D., Prade, H., (eds.) Fundamentals of Fuzzy Sets. The Handbooks of Fuzzy Sets Series, vol.1, pp. 125–193. Kluwer (2000)
36. Grabisch, M.: On equivalence classes of fuzzy connectives: the case of fuzzy integrals. IEEE Trans. Fuzzy Syst. **3**, 96–109 (1995)
37. Grabisch, M.: Fuzzy integral in multicriteria decision making. Fuzzy Sets Syst. **69**, 279–298 (1996)
38. Murofushi, T., Sugeno, M.: Some quantities represented by the Choquet integral. Fuzzy Sets Syst. **56**, 229–235 (1993)

Entropy Measures and Views of Information

Bernadette Bouchon-Meunier and Christophe Marsala

Abstract Among the countless papers written by Ronald R. Yager, those on Entropies and measures of information are considered, keeping in mind the notion of view of a set, in order to point out a similarity between the quantities introduced in various frameworks to evaluate a kind of entropy. We define the concept of entropy measure and we show that its main characteristic is a form of monotonicity, satisfied by quantities scrutinised by R.R. Yager.

1 Introduction

Entropies and measures of information, fuzziness, specificity are key concepts in the management of data, knowledge and information. R.R. Yager has pointed out their importance from the early beginning of his work on fuzzy systems and soft computing. It is no coincidence that he co-founded the International Conference on Information Processing and Management of Uncertainty in 1986, his area of interest being clearly at the cross roads of both domains of uncertain knowledge representation and information evaluation in digital media. Among the long list of his publications, those related to quantities measuring information, entropy, measures of fuzziness, constitute an important mine of tools to capture the uncertainty inherent in data and knowledge management. We focus on some of these quantities and we show that they are of different types, depending on their properties and the purpose of their introduction. The same word of *entropy* has been used with different meanings and we propose to consider properties enabling the user to choose an appropriate entropy measure for a given problem.

B. Bouchon-Meunier (✉) · C. Marsala
Sorbonne Universités, UPMC Univ Paris 06, CNRS, LIP6 UMR 7606,
4 Place Jussieu, 75005 Paris, France
e-mail: Bernadette.Bouchon-Meunier@lip6.fr

C. Marsala
e-mail: Christophe.Marsala@lip6.fr

© Springer International Publishing Switzerland 2017
J. Kacprzyk et al. (eds.), *Granular, Soft and Fuzzy Approaches
for Intelligent Systems*, Studies in Fuzziness and Soft Computing 344,
DOI 10.1007/978-3-319-40314-4_3

47

In Sect. 2, we introduce information as a view of a set and we consider solutions to evaluate views. In Sect. 3, we define entropy measures and their associated possible properties of symmetry, ψ-recursivity, monotonicity and additivity. Section 4 presents classic examples of entropy measures, which leads to considering weakened properties of recursivity and additivity, regarded as new forms of monotonicity. We then consider these properties in the case of entropy measures introduced by R.R. Yager, based on aggregation operators as detailed in Sect. 5, in the framework of theory of evidence as described in Sect. 6, and in association with similarity relations as presented in Sect. 7. In Sect. 8, we tackle entropy measures dealing with intuitionistic fuzzy sets, which have attracted the interest of R.R. Yager, but not in what concerns entropy or measures of information. We then conclude.

2 Information and Measure of Information

2.1 Information Is a View on a Set

Nowadays, measuring information is a very hot topic and one of the main challenges is to define the concept of information, as it is done in the recent work by Lotfi Zadeh [22]. In this paper, he considers two aspects of information: probabilistic, and possibilistic, and their combination.

In this paper, we do not focus on defining information but we discuss on the measurement of information and the evaluation of a measure of information.

Let us consider a set of objects (or, in particular cases, events) that represent a physical representation of the real world. In this paper, for the sake of clarity, only finite sets are considered but this work could be generalised to non-countable sets. We consider a σ-algebra B defined on a finite universe U. In order to formally manipulate, or to make predictions, or to evaluate a subset X of B, a formal representation of X, a *measure*, is usually used that identifies objects $x \in X$ through a given *view*. In Kampé de Fériet's work, views are *observers* of the objects [10, 12].

In this paper, a view is a measure that reports a kind of information on the objects of X. This measure is dependent on the particular objective that should be fulfilled or the aim of the user. Typically:

Definition 1 A *view* is a mapping from 2^X to $\mathbb{R}^+$.

For instance, in order to make a prediction, we are interested in obtaining some information about the occurrence of the objects, and the classical view in this case is to use a probability of the occurrence of each object x from X. Another example, in order to define a vague category composed of objects, a useful view is a membership function that defines this category as a fuzzy set of X.

2.2 Evaluation of a View

Evaluating a view is an important way to gain more knowledge on X with regard to the considered view. Such an evaluation is obtained by considering all the values of the view in order to obtain a single value that expresses a global knowledge on the view. Aggregating all the values is thus done by means of a *measure of information*. In the literature, measures of information take various forms and we give the most basic hereunder.

2.2.1 Fuzzy Measure

Depending on the information that should be handled through the view, the measure is associated with specific properties: monotonicity, maximality, minimality, etc. Classically, when related to the occurrence of the objects, a view is a probability. In fuzzy sets theory, a view is often a fuzzy measure.

Definition 2 A *fuzzy measure* is a mapping $f : 2^X \to [0, 1]$ such that

 (i) $f(\emptyset) = 0$,
 (ii) $f(X) = 1$,
 (iii) $\forall A, B \in 2^X$, if $A \subseteq B$ then $f(A) \leq f(B)$.

But a view could be any kind of measure. For instance, a view on X could be defined as the association of a weight to each element of X. In evidence theory, mass assignment is an example of such an association. But weights could be of any kind (the age of each element x_i, a given price, a duration, etc.).

Usually, a view enables us to help the *ranking* of elements from X in order to select one, for instance, the most probable event from X, or the most representative subset set of X.

2.2.2 Probabilistic Case

There exists several measures of information to value the global information brought out by a probability distribution. The well-known Shannon entropy is such a measure of information enabling the evaluation of a particular view. It is usually referred to as a measure of the probabilistic disorder of the set X.

Definition 3 Let a σ-algebra B be defined on a finite universe U, and let $X_n = \{x_1, \dots, x_n\}$ for $n \in \mathbb{N}$. Let $p : B \to [0, 1]$, be a probability distribution over X_n, p_i being associated with x_i. The *Shannon entropy* of p is defined as:

$$H(p) = -\sum_{i=1}^{n} p_i \log p_i$$

Here, the view on X_n is the probability p and the Shannon entropy $H(p)$ offers a value related to p such that [1]:

1. $H(p) \geq 0$ for any probability distribution p *(non negativity)*
2. $H(p) = 0$ if $\exists k \in \{1, \ldots, n\}, p_k = 1$ and $\forall i \neq k, p_i = 0$ *(minimality)*
3. $H(p) = 1$ if $p = (\frac{1}{n}, \ldots, \frac{1}{n})$ *(normality)*
4. $H(pq) = H(p) + H(q)$ for any probability distributions p and q *(additivity)*

However, this kind of evaluation of the view is highly focused on the probabilistic information based on p. In particular, the additivity property is a strong commitment for the evaluation of the view.

In his work, J. Kampé de Fériet introduced a new way of aggregating information by considering any operation of composition to construct $H(pq)$ from $H(p)$ and $H(q)$ [11].

2.2.3 Fuzzy Case

In some cases, the view is not a probability, but for instance a membership function defining a fuzzy set.

For instance, De Luca and Termini have proposed an *entropy measure of a fuzzy set* defined by its membership function $\mu : X \to [0, 1]$ as follows [7]:

1. $H(\mu) = 0$ if and only if μ takes values 0 or 1
2. $H(\mu)$ is maximum if and only if μ assumes always the value $\frac{1}{2}$
3. $H(\mu) \geq H(\mu^*)$ for any μ^* sharpened version of μ, i.e. any fuzzy set such that $\mu(x) \leq \mu^*(x)$ if $\mu(x) \geq \frac{1}{2}$ and $\mu(x) > \mu^*(x)$ if $\mu(x) \leq \frac{1}{2}$

2.2.4 Disorder and Homogeneity

Similarities between properties of measures of information in the probabilistic case and the fuzzy case are evident as both enable the aggregation of a view on X.

As previously said, Shannon entropy is usually considered as a measure of the disorder existing when a prediction of the occurrence of one element of X should be done. The higher the entropy, the less predictable the event x from X to occur.

However, another interpretation could be used, that refers to the physical interpretation of the Bolztmann entropy. In physics and thermodynamics, the entropy of a system grows over time and is maximum when all the studied particles have the same temperature.

Thus, we can view an entropy as *a measure of homogeneity* of the set X through the considered view.

It is easy to see that existing measures aim at evaluating homogeneity:

- Shannon entropy is maximum when all probabilities are equal
- De Luca and Termini entropy is maximum when all the elements of X have the same membership degree of $\frac{1}{2}$ that refers to an incomplete membership to the fuzzy set.

3 Entropy Measures

The various evaluations of a view we have mentioned have very different properties, even though they intuitively belong to the same family. We propose a formal approach of this family, introducing the concept of entropy measure which encompasses the previous quantities, pointing out major properties they may share. We then study particular frameworks where such properties have been considered, pointing out that some quantities introduced by R.R. Yager with the purpose of evaluating views in specific environments such as evidence theory or similarities, as well as the intuitionistic paradigm he was interested in, have mainly in common a general property of monotonicity with regard to a kind of refinement of information. We will then conclude that such a form of monotonicity is the most important property shared by those very different entropy measures, the concept of refinement of information being strongly dependent on the framework.

Starting from the seminal paper by Aczél and Daróczy on the so-called inset entropy [2], we consider again a σ-algebra B defined on a finite universe U.

For any integer n, we denote: $X_n = \{(x_1, \ldots, x_n) | x_i \in B, \forall i = 1, \ldots, n\}$, $P_n = \{(p_1, \ldots, p_n) | p_i \in [0, 1]\}$, p_i being associated with x_i through a function $p : B \to [0, 1]$, a particular case being a probability distribution defined on (U, B), $W_n = \{(w_{x_1}, \ldots, w_{x_n}) | w_{x_i} \in \mathbb{R}^+, \forall i = 1, \ldots, n\}$, a family of n-tuples of weights[1] associated with n-tuples of B through a function $f : B \to \mathbb{R}^{+n}$, such that $f(x_1, \ldots, x_n) = (w_{x_1}, \ldots, w_{x_n})$.

An **entropy measure** is a sequence of mappings $E_n : X_n \times P_n \times W_n \to \mathbb{R}^+$ satisfying several properties among the following ones.

3.1 Symmetry

For any permutation σ of $1, \ldots, n$, we have:

$$E_n \begin{pmatrix} x_1, & \ldots, & x_n \\ p_1, & \ldots, & p_n \\ w_{x_1}, & \ldots, & w_{x_n} \end{pmatrix} = E_n \begin{pmatrix} x_{\sigma(1)}, & \ldots, & x_{\sigma(n)} \\ p_{\sigma(1)}, & \ldots, & p_{\sigma(n)} \\ w_{x_{\sigma(1)}}, & \ldots, & w_{x_{\sigma(n)}} \end{pmatrix}$$

---

[1] In the following, for the sake of simplicity, w_{x_i} will be denoted w_i when the meaning of i is clear.

3.2 ψ-Recursivity

$$E_n \begin{pmatrix} x_1, & x_2, & \dots, & x_n \\ p_1, & p_2, & \dots, & p_n \\ w_{x_1}, & w_{x_2} & \dots, & w_{x_n} \end{pmatrix} =$$

$$E_{n-1} \begin{pmatrix} x_1 \cup x_2, & x_3 & \dots, & x_n \\ p_1 + p_2, & p_3 & \dots, & p_n \\ w_{x_1 \cup x_2}, & w_{x_3} & \dots, & w_{x_n} \end{pmatrix} + \psi \begin{pmatrix} x_1, & x_2 \\ p_1, & p_2 \\ w_{x_1}, & w_{x_2} \end{pmatrix} E_2 \begin{pmatrix} x_1, & x_2 \\ \frac{p_1}{p_1 + p_2}, & \frac{p_2}{p_1 + p_2} \\ w_{x_1}, & w_{x_2} \end{pmatrix} \quad (1)$$

for a function $\psi : X_2 \times P_2 \times W_2 \to \mathbb{R}^+$.

The classic property of recursivity corresponds to:

$$\psi_0 \begin{pmatrix} x_1, & x_2 \\ p_1, & p_2 \\ w_{x_1}, & w_{x_2} \end{pmatrix} = p_1 + p_2,$$

weights not being taken into account.

3.3 Monotonicity

$$E_n \begin{pmatrix} x_1, & \dots, & x_n \\ p_1, & \dots, & p_n \\ w_{x_1}, & \dots, & w_{x_n} \end{pmatrix} \leq E_n \begin{pmatrix} x'_1, & \dots, & x'_n \\ p'_1, & \dots, & p'_n \\ w_{x'_1}, & \dots, & w_{x'_n} \end{pmatrix} \text{ if } \begin{pmatrix} x_1, & \dots, & x_n \\ p_1, & \dots, & p_n \\ w_{x_1}, & \dots, & w_{x_n} \end{pmatrix} \prec \begin{pmatrix} x'_1, & \dots, & x'_n \\ p'_1, & \dots, & p'_n \\ w_{x'_1}, & \dots, & w_{x'_n} \end{pmatrix}$$

for some partial order $\prec$ on $\bigcup_n X_n \times P_n \times W_n$.

Examples of monotonicity can be based on the following partial orders:

$$\begin{pmatrix} x_1, & \dots, & x_n \\ p_1, & \dots, & p_n \\ w_{x_1}, & \dots, & w_{x_n} \end{pmatrix} \prec \begin{pmatrix} x_1, & \dots, & x_n \\ p_1, & \dots, & p_n \\ w'_{x_1}, & \dots, & w'_{x_n} \end{pmatrix}$$

if and only if, $\forall i$ the following conditions (M1) or (M2) are satisfied:

(M1) • $w'_i \leq w_i \Leftrightarrow w'_i \geq \frac{1}{2}$;
        • $w'_i \geq w_i \Leftrightarrow w'_i < \frac{1}{2}$.
(M2) $w'_i \geq w_i$.

Other examples will be studied later.

3.4 Additivity

Let us further consider a second finite universe U', a σ-algebra B' on U'. In a way similar to the situation on U, we consider

- $X'_m = \{(x'_1, \dots, x'_m) | x'_i \in B', \forall i\}$;
- $P'_m = \{(p'_1, \dots, p'_m) | p'_i \in [0, 1]\}, p'_i$ being associated with x'_i through a function p' : $B' \to [0, 1]$;
- $W'_m = \{(w'_1, \dots, w'_m) | w'_i \in \mathbb{R}^+; \forall i\}$, a family of m-tuples of weights associated with m-tuples of elements of B' through a function $f' : B' \to \mathbb{R}^+$, such that $f'(x'_i) = w'_i$.

We further suppose that there exist two combination operators $\star$ and $\circ$ enabling us to equip the Cartesian product of U with similar distributions:

- $P \star P'_{n \times m} = \{(p_1 \star p'_1, \dots, p_i \star p'_j, \dots) | p_i \star p'_j \in [0, 1]\}$, $p_i \star p'_j$ being associated with (x_i, x'_j) for any i and j through a function $p \star p'$,
- $W \circ W'_{n \times m} = \{(w_{1,1}, \dots, w_{i,j}, \dots) | w_{i,j} \in \mathbb{R}^+, \forall i, j\}$, is defined through a function $f \circ f' : B \times B' \to \mathbb{R}^+$, such that: $f \circ f'(x_1, x'_j) = w_{i,j}$ for all $i = 1, \dots, n$ and $j = 1, \dots, m$.

The classic additivity property stands in the case where U and U' are independent universes, p and p' being probability distributions on (U, B) and (U', B'), weights generally not being taken into account. It yields:

$$E_{n \times m} \begin{pmatrix} (x_1, x'_1), & (x_1, x'_2), & \dots, & (x_i, x'_j), & \dots, & (x_n, x'_m) \\ p_1 \star p'_1, & p_1 \star p'_2, & \dots, & p_i \star p'_j, & \dots, & p_n \star p'_m \\ w_{x_1, x'_1}, & w_{x_1, x'_2}, & \dots, & w_{x_i, x'_j}, & \dots, & w_{x_n, x'_m} \end{pmatrix} =$$

$$E_n \begin{pmatrix} x_1, & \dots, & x_n \\ p_1, & \dots, & p_n \\ w_{x_1}, & \dots, & w_{x_n} \end{pmatrix} + E_m \begin{pmatrix} x'_1, & \dots, & x'_m \\ p'_1, & \dots, & p'_m \\ w_{x'_1}, & \dots, & w_{x'_n} \end{pmatrix} \quad (2)$$

4 Classic Examples of Entropy Measures

A particular case of entropy measures is the general class of **inset entropies** introduced by Aczél and Daróczy [2], who restricted their study to $X_n = \{(x_1, \dots, x_n) | x_i \in B, \forall i$ such that $x_i \cap x_j = \emptyset, \forall j\}$ and considered sequences of mappings independent of $W_n \forall n: I_n : X_n \times P_n$, where elements of P_n are probabilities. They consider such mappings satisfying the ψ_0-recursivity as well as the symmetry and measurability with regard to the probability. They have proved that I_n is necessarily of the form:

$$I_n \begin{pmatrix} x_1, x_2 \\ p_1, p_2 \end{pmatrix} = g(\bigcup_i x_i) - \sum_i p_i g(x_i) - K \sum_i p_i \log p_i$$

for a constant K and a function $g : B \to \mathbb{R}$

A **possibilistic inset entropy** has then been introduced in [3] in the case where f lies in $[0, 1]$ and $\max_i w_i = 1$. It can be proved that a possibilistic inset entropy satisfies properties of symmetry, measurability and ψ_0-recursivity if and only if:

$$E_n \begin{pmatrix} x_1, & x_2, & \ldots, & x_n \\ p_1, & p_2, & \ldots, & p_n \\ w_1, & w_2, & \ldots, & w_n \end{pmatrix} = G(\max_i w_i) - \sum_i p_i G(w_i) - K \sum_i p_i \log p_i$$

for a constant K and a function $G : [0, 1] \to \mathbb{R}^+$, assuming that $w_{x_i \cup x_j} = \max(w_{x_i}, w_{x_j})$, $\forall i, j$.

One of the seminal papers on entropies involving elements different from probabilities introduces [8] a **weighted entropy** as follows [8]:

$$I_n^w \begin{pmatrix} x_1, & x_2, & \ldots, & x_n \\ p_1, & p_2, & \ldots, & p_n \\ w_1, & w_2, & \ldots, & w_n \end{pmatrix} = - \sum_i w_i p_i \log p_i.$$

This quantity is proportional to Shannon entropy $H_n(p_1, \ldots, p_n)$ when all weights w_i are equal. It is monotonic with regard to an order on the weights defined as follows:

$$\begin{pmatrix} x_1, & x_2, & \ldots, & x_n \\ p_1, & p_2, & \ldots, & p_n \\ w_1, & w_2, & \ldots, & w_n \end{pmatrix} \prec \begin{pmatrix} x_1, & x_2, & \ldots, & x_n \\ p_1, & p_2, & \ldots, & p_n \\ w_1', & w_2', & \ldots, & w_n' \end{pmatrix}$$

with $w_i \le w_i'$ for all i. It is also ψ_0-recursive:

$$E_n \begin{pmatrix} x_1, & x_2, & \ldots, & x_n \\ p_1, & p_2, & \ldots, & p_n \\ w_1, & w_2, & \ldots, & w_n \end{pmatrix} =$$

$$E_{n-1} \begin{pmatrix} x_1 \cup x_2, & x_3, & \ldots, & x_n \\ p_1 + p_2 & p_3, & \ldots, & p_n \\ f_2(x_1 \cup x_2), & w_3, & \ldots, & w_n \end{pmatrix} + (p_1 + p_2) \times E_2 \begin{pmatrix} x_1, & x_2 \\ \frac{p_1}{p_1+p_2}, & \frac{p_2}{p_1+p_2} \\ w_1, & w_2 \end{pmatrix} \qquad (3)$$

with $f_2(x_1 \cup x_2) = \frac{(p_1 w_1 + p_2 w_2)}{(p_1 + p_2)}$, all other weights being unchanged.

As indicated in Sect. 2.2.3, another classic entropy measure has been introduced by De Luca and Termini [7] as a measure of fuzziness, in the case where f is the membership function of a fuzzy set on U:

$$E_n^{DLT} \begin{pmatrix} x_1, & x_2, & \ldots, & x_n \\ p_1, & p_2, & \ldots, & p_n \\ w_1, & w_2, & \ldots, & w_n \end{pmatrix} = - \sum_i w_i \log w_i - \sum_i (1 - w_i) \log(1 - w_i).$$

A major property of this quantity is its monotonicity with respect to the above mentioned partial order *(M1)*, called *sharpness*.

Let us remark that De Luca and Termini have also introduced the total entropy:

$$E_n^{TE} \begin{pmatrix} x_1, \ \ldots, \ x_n \\ p_1, \ \ldots, \ p_n \\ w_1, \ \ldots, \ w_n \end{pmatrix} = -\sum_i p_i \log p_i - \sum_i p_i \big(w_i \log w_i - (1 - w_i) \log(1 - w_i) \big).$$

taking into account probabilities p_i and degrees of fuzziness w_i associated with elements x_i, $i = 1, \ldots n$ of U. It can be proved [3] that, in the case where $\max_i w_i = 1$, the weights representing a normal possibility distribution, the total entropy is a possibilistic ψ_0-recursive inset entropy, with $K = 1$ and $G(x) = G(1) - x \log x - (1 - x) \log(1 - x)$, $\forall x \in [0, 1]$.

It appears that existing entropies or measures of fuzziness do not have the same behaviour with regard to the major properties of ψ-recursivity, additivity or monotonicity. Nevertheless, as pointed out by Ronald Yager in several of his papers [17, 18], these quantities are roughly in the same category. An approach to support such a claim is to observe that they evaluate an amount of information or a decrease of uncertainty after a given observation, as mentioned by Renyi [15].

Therefore, we suggest to consider a common feature between all these quantities existing in the literature under the name of entropy or taking a form similar to classic entropies, the best known being obviously the Shannon entropy. This common factor is the fact that a refinement of the tool to perform observations increases the entropy measure. This approach can be compared to Mugur-Schächter's works on the general relativity of descriptions, considering that any process of knowledge extraction is associated with epistemic operators called a delimiter and a view, representing the influence of the context and the observation tool on the available information [14].

- Such a refinement can be obtained by means of a partial order on B, inducing a **monotonicity property** as presented above.
- Another kind of refinement can be associated with a decrease of the coarseness of a partition of the universe of discourse, and this corresponds to a property of **weak recursivity** defined as follows:

$$(M3) \qquad E_n \begin{pmatrix} x_1, \ x_2, \ \ldots, \ x_n \\ p_1, \ p_2, \ \ldots, \ p_n \\ w_1, \ w_2, \ \ldots, \ w_n \end{pmatrix} \geq E_{n-1} \begin{pmatrix} x_1 \cup x_2, \ x_3, \ \ldots, \ x_n \\ p_1 + p_2, \ p_3, \ \ldots, \ p_n \\ w_{x_1 \cup x_2}, \ w_3, \ \ldots, \ w_n \end{pmatrix}$$

- Another means to refine information is to consider a secondary universe of discourse providing more details on the observed phenomenon. Such a refinement leads to a property of **weak additivity** stating the following:

$$(M4) \qquad E_{n \times m} \begin{pmatrix} (x_1, x'_1), & \cdots, & (x_n, x'_m) \\ p_1 \star p'_1, & \cdots, & p_n \star p'_m \\ w_1 \text{o} w'_1, & w_1 \text{o} w'_2, & \cdots, & w_n \text{o} w'_m \end{pmatrix} \geq$$

$$\max \left[E_n \begin{pmatrix} x_1, & \cdots, & x_n \\ p_1, & \cdots, & p_n \\ w_1, & \cdots, & w_n \end{pmatrix}, E_m \begin{pmatrix} x_1, & \cdots, & x_m \\ p_1, & \cdots, & p_m \\ w'_1, & \cdots, & w'_m \end{pmatrix} \right]$$

Coming back to the measure of fuzziness proposed by De Luca and Termini, it can be observed that, in the case where the weights are possibility degrees, it is also weakly recursive and then monotonic with regard to the refinement of granular information:

$$E_n^{DLT} \begin{pmatrix} x_1, & x_2, & \cdots, & x_n \\ p_1, & p_2, & \cdots, & p_n \\ w_1, & w_2, & \cdots, & w_n \end{pmatrix} \geq E_{n-1}^{DLT} \begin{pmatrix} x_1 \cup x_2, & x_3, & \cdots, & x_n \\ p_1 + p_2, & p_3, & \cdots, & p_n \\ \max(w_1, w_2), & w_3, & \cdots, & w_n \end{pmatrix}$$

Let us remark that any entropy measure satisfying a property of ψ-recursivity is also weakly recursive. If it satisfies the additivity property, it is also weakly additive.

5 Measures of Entropy Related to Aggregation Operators

Yager [18] considers measures of entropy based on OWA operators involving argument-dependent weights. For a parameter $\alpha \geq 1$, he considers:

$$E_n^{Y,\alpha} \begin{pmatrix} x_1, & \cdots, & x_n \\ p_1, & \cdots, & p_n \\ w_1, & \cdots, & w_n \end{pmatrix} = 1 - \sum_i w_i p_i.$$

with probabilities p_i and weights $w_i = \dfrac{p_i^\alpha}{\sum_i p_i^\alpha}$, $\forall i = 1, \ldots, n$. We can observe that such quantities have basic properties similar to Shannon entropy, such as symmetry and continuity. Nevertheless, no property of ψ-recursivity can be proved in the general case. The general form:

$$E_n^{Y,\alpha} \begin{pmatrix} x_1, & \cdots, & x_n \\ p_1, & \cdots, & p_n \\ w_1, & \cdots, & w_n \end{pmatrix} = 1 - \sum_i \frac{p_i^{\alpha+1}}{\sum_i p_i^\alpha}.$$

reduces to:

$$E_n^{Y,1} \begin{pmatrix} x_1, & \cdots, & x_n \\ p_1, & \cdots, & p_n \\ w_1, & \cdots, & w_n \end{pmatrix} = 1 - \sum_i p_i^2.$$

in the case where $\alpha = 1$, which equals

$$\frac{1}{2} E_n^{D,2} \begin{pmatrix} x_1, & \ldots, & x_n \\ p_1, & \ldots, & p_n \\ w_1, & \ldots, & w_n \end{pmatrix},$$

where $E_n^{D,2}$ denotes the Daroczy entropy of type 2 [6]. Therefore, $E_n^{Y,1}$ is ψ_2-recursive, with:

$$\psi_2 \begin{pmatrix} x_1, & x_2 \\ p_1, & p_2 \\ p_1, & p_2 \end{pmatrix} = (p_1 + p_2)^2.$$

We deduce that $E_n^{Y,1}$ is monotonic with regard to the refinement of information, as follows:

$$E_n^{Y,1} \begin{pmatrix} x_1, & x_2, & \ldots, & x_n \\ p_1, & p_2, & \ldots, & p_n \\ p_1, & p_2, & \ldots, & p_n \end{pmatrix} \geq E_{n-1}^{Y,1} \begin{pmatrix} x_1 \cup x_2, & x_3, & \ldots, & x_n \\ p_1 + p_2, & p_3, & \ldots, & p_n \\ p_1 + p_2, & p_3, & \ldots, & p_n \end{pmatrix}.$$

An analogous form of entropy measure is also investigated [18] as:

$$E_n^{Y,p} \begin{pmatrix} x_1, & \ldots, & x_n \\ p_1, & \ldots, & p_n \\ w_1, & \ldots, & w_n \end{pmatrix} = -\sum_i w_i(1 - w_i)$$

where w_i is a membership degree associated with $x_i, i = 1, \ldots, n$. It is clear that it is the same form as $E_n^{Y,1}$ when weights are probabilities. In its general form, $E_n^{Y,p}$ is a measure of fuzziness and satisfies monotonicity (M1).

6 Framework of Theory of Evidence

We consider a frame of evidence U, with a basic assignment $m : 2^U \to [0, 1]$, such that $m(\emptyset) = 0$ and $\sum_{x \in 2^U} m(x) = 1$. Yager introduced [19, 20] the entropy associated with this basic assignment as follows, for the family $(x_1, \ldots, x_n)$ of focal elements associated with m:

$$E_n^Y \begin{pmatrix} x_1, & \ldots, & x_n \\ m_1, & \ldots, & m_n \\ w_1, & \ldots, & w_n \end{pmatrix} = -\sum_i m_i \log w_i$$

with $m_i = m(x_i)$ and $w_i = Pl(x_i) = \sum_{\{j \mid x_j \cap x_i \neq \emptyset\}} m_j$ being the plausibility of x_i, for all $i = 1, 2 \ldots$.

It is easy to see that E_n^Y satisfies the weak recursivity property:

$$E_n^Y \begin{pmatrix} x_1, & x_2, & \ldots, & x_n \\ m_1, & m_2, & \ldots, & m_n \\ w_1, & w_2, & \ldots, & w_n \end{pmatrix} \geq E_{n-1}^Y \begin{pmatrix} x_1 \cup x_2, & x_3, & \ldots, & x_n \\ m_1 + m_2, & m_3, & \ldots, & m_n \\ w_{x_1 \cup x_2}, & w_3, & \ldots & w_n \end{pmatrix}$$

Proof: We note $\chi_{x_\alpha, x_\beta} = 1$ if $x_\alpha \cap x_\beta \neq \emptyset$, and $\chi_{x_\alpha, x_\beta} = 0$ if $x_\alpha \cap x_\beta = \emptyset$.
We have:

$$\Delta_{n-1} = E_{n-1}^Y \begin{pmatrix} x_1 \cup x_2, & x_3, & \ldots, & x_n \\ m_1 + m_2, & m_3, & \ldots, & m_n \\ w_{x_1 \cup x_2}, & w_3, & \ldots & w_n \end{pmatrix} =$$

$$- \sum_{i=3,\ldots,n} m_i \log \left(\sum_{j=3,\ldots,n} \chi_{x_i, x_j} m_j + \chi_{x_i, x_1 \cup x_2} (m_1 + m_2) \right)$$

$$- (m_1 + m_2) \log \left(\sum_{j=3,\ldots,n} \chi_{x_1 \cup x_2, x_j} m_j + m_1 + m_2 \right). \qquad (4)$$

Let us remark that: $\chi_{x_i, x_1 \cup x_2}(m_1 + m_2) \geq \chi_{x_i, x_1} m_1 + \chi_{x_i, x_2} m_2$ for any $i = 3, \ldots, n$ and $\chi_{x_i, x_1 \cup x_2} m_1 \geq \chi_{x_i, x_1} m_1$ for any $i = 3, \ldots, n$.
We get:

$$\Delta_{n-1} \leq - \sum_{i=3,\ldots,n} m_i \log \left(\sum_{j=3,\ldots,n} \chi_{x_i, x_j} m_j + \chi_{x_i, x_1} m_1 + \chi_{x_i, x_2} m_2 \right)$$

$$- m_1 \log \left(\sum_{j=3,\ldots,n} \chi_{x_j, x_1} m_j + \chi_{x_1, x_1} m_1 + \chi_{x_1, x_2} m_2 \right)$$

$$- m_2 \log \left(\sum_{j=3,\ldots,n} \chi_{x_j, x_2} m_j + \chi_{x_2, x_2} m_2 + \chi_{x_1, x_2} m_1 \right)$$

and thus

$$\Delta_{n-1} \leq - \sum_{i=1,\ldots,n} m_i \log \left(\sum_{i=1,\ldots,n} \chi_{x_i, x_j} m_j \right).$$

We deduce that:

$$\Delta_{n-1} \leq E_n^Y \begin{pmatrix} x_1, & \ldots, & x_n \\ m_1, & \ldots, & m_n \\ w_1, & \ldots, & w_n \end{pmatrix}$$

7 Entropy Measures Under Similarity Relations

Following Yager's suggestion [17], let us consider a similarity relation S on $U = \{x_1, \ldots, x_n\}$, reflexive, symmetric and min-transitive and let us define the entropy:

$$E_n \begin{pmatrix} x_1, x_2, \ldots, x_n \\ p_1, p_2, \ldots, p_n \\ \overline{S}_1, \overline{S}_2, \ldots, \overline{S}_n \end{pmatrix} = -\sum_{x_i \in U} p_i \log \overline{S}_i$$

with $\overline{S}_i = \sum_{x_j \in U} p_j S(x_i, x_j)$ for all $i = 1, \ldots n$.

This quantity is an entropy measure satisfying the monotonicity property with respect to the order on similarities S and S' defined as follows:

$$S \prec S' \Leftrightarrow S(x_i, x_j) \le S'(x_i, x_j) \; \forall i, j.$$

This order provides an order on

$$\begin{pmatrix} x_1, \ldots, x_n \\ p_1, \ldots, p_n \\ \overline{S}_1, \ldots, \overline{S}_n \end{pmatrix} \le \begin{pmatrix} x_1, \ldots, x_n \\ p_1, \ldots, p_n \\ \overline{S}'_1, \ldots, \overline{S}'_n \end{pmatrix}$$

with $\overline{S}'_i = \sum_{x_j \in U} p_j S'(x_i, x_j)$.

Another monotonicity property of this entropy measure is the weak additivity (M4) [3], obtained by defining a joint similarity relation $S \times S'$ on the Cartesian product $U \times U'$ as follows, for two similarity relations S defined on U and S' defined on U':

$$S \times S'((x_i, y_j), (x_k, y_l)) = \min \left(S(x_i, x_k), S'(y_j, y_l) \right)$$

for any x_i and x_k in U, any y_j and y_l in U'.

8 Intuitionistic Quantities

First of all, some basic recalls on intuitionistic fuzzy sets are presented here. Let X be a universe, an *intuitionistic fuzzy set* (IFS) A of X is defined by:

$$A = \{(x, \mu_A(x), v_A(x)) | x \in X\}$$

with $\mu : X \to [0, 1]$, $v : X \to [0, 1]$ and $0 \le \mu_A(x) + v_A(x) \le 1$, $\forall x \in X$. Here, $\mu_A(x)$ and $v_A(x)$ represent respectively the membership degree and the non-membership degree of x in A.

Given an intuitionistic fuzzy set A of X, the *intuitionistic index of x to A* is defined for all $x \in X$ as: $\pi_A(x) = 1 - (\mu_A(x) + v_A(x))$. This index represents the hesitancy lying on the membership of x in A.

There exists several works in IFS theory that propose an *entropy of an intuitionistic fuzzy set A*. In order to summarise their form, we need to extend our presented model (Sect. 3) to complex numbers by considering that weights belong to $Z_n = \{(z_{x_1}, \ldots, z_{x_n}) \mid z_{x_i} \in \mathbb{C}, \forall i = 1, \ldots, n\}$. It is justified by the fact that each $x \in X$ being associated with two values $\mu_A(x), \nu_A(x)$, it could thus be associated with a point in a two dimensional space. It is a classical representation in intuitionistic works even if a 3D-representation could be used (as, for instance, in [16]).

Here, we define for each $x \in X$, $z_A(x) = \mu_A(x) + i\nu_A(x)$, considering that $\mu_A(x)$ is the real part of z_x and $\nu_A(x)$ is its imaginary part.

Entropy of intuitionistic fuzzy sets are summarised as:

$$E_n^{IFS} \begin{pmatrix} x_1, & \ldots, & x_n \\ p_1, & \ldots, & p_n \\ z_A(x_1) & \ldots & z_A(x_n) \end{pmatrix}.$$

For instance, the entropy defined in [16] is rewritten as:

$$E_n^{S} \begin{pmatrix} x_1, & \ldots, & x_n \\ p_1, & \ldots, & p_n \\ z_A(x_1), & \ldots, & z_A(x_n) \end{pmatrix} = 1 - \frac{1}{2n} \sum_{i=1}^{n} |\mu_A(x_i) - \nu_A(x_i)|$$

with $\mu_A(x) = \Re(z_A(x))$ and $\nu_A(x) = \Im(z_A(x))$ for all $x \in X$.

It could be interesting to see that E_n^S is related to $E_n^{Y,\alpha}$ in the particular case where $\Re(z_A(x)) = \frac{p_i^{\alpha+1}}{\sum_i p_i^\alpha}$ and $\Im(z_A(x)) = 0$ for all $x \in X$. In this case, $E_n^S = 1 - \frac{1}{2n} + \frac{1}{2n} E_n^{Y,\alpha}$ showing that E_n^S and $E_n^{Y,\alpha}$ are linearly dependent.

Various definitions are recalled in [9]. In this paper, it can be seen that the monotonicity property is ensured by definition. The authors present several definitions that lie on the definition of a partial order on W_n and the concept of *less fuzzy than*. For instance, we recall here two definitions that are used:

(M5)
$$E_n^{IFS}(A) \leq E_n^{IFS}(B) \quad \text{if } A \text{ is } less\,fuzzy \text{ than } B$$
i.e. $\mu_A(x) \leq \mu_B(x)$ and $\nu_A(x) \geq \nu_B(x)$ for $\mu_B(x) \leq \nu_B(x), \forall x \in X$,
or $\mu_A(x) \geq \mu_B(x)$ and $\nu_A(x) \leq \nu_B(x)$ for $\mu_B(x) \geq \nu_B(x), \forall x \in X$

and

(M6)
$$E_n^{IFS}(A) \leq E_n^{IFS}(B) \quad \text{if } A \text{ is } less\,fuzzy \text{ than } B$$
i.e. $A \subseteq B$ for $\mu_B(x) \leq \nu_B(x), \forall x \in X$,
or $B \subseteq A$ for $\mu_B(x) \geq \nu_B(x), \forall x \in X$.

Each definition of the monotonicity produces the definition of a particular form of E^{IFS}. Thus, E_n^S satisfies (M5).

In [9], the following entropy that satisfies (M6) is introduced:

$$E_n^G \begin{pmatrix} x_1, & \cdots, & x_n \\ p_1, & \cdots, & p_n \\ z_A(x_1), & \cdots, & z_A(x_n) \end{pmatrix} = \frac{1}{2n} \sum_{i=1}^{n} \left(1 - |\mu_A(x_i) - \nu_A(x_i)|\right)(1 + \pi_A(x_i))$$

with, $\forall x \in X$, $\mu_A(x) = \Re(z_A(x))$, $\nu_A(x) = \Im(z_A(x))$ and $\pi_A(x) = 1 - \Re(z_A(x)) - \Im(z_A(x))$.

Another way to define entropy is presented in [5] where the definition is based on extensions of the Hamming distance and the Euclidian distance to intuitionistic fuzzy sets. For instance, the following entropy is proposed:

$$E_n^B \begin{pmatrix} x_1, & \cdots, & x_n \\ p_1, & \cdots, & p_n \\ z_A(x_1), & \cdots, & z_A(x_n) \end{pmatrix} = \sum_{i=1}^{n} \pi_A(x_i).$$

with $\pi_A(x) = 1 - \Re(z_A(x)) - \Im(z_A(x))$ for all $x \in X$.

As stated in [5], this definition satisfies the following (M7) property:

(M7) $E_n^{IFS}(A) \leq E_n^{IFS}(B)$ if $A \leq B$
  i.e. $\mu_A(x) \leq \mu_B(x)$ and $\nu_A(x) \leq \nu_B(x), \forall x \in X$,

Further work could be found in Yager's paper [21]. In this paper, R.R. Yager has expressed his interest in IFS in which he studies the concept of specificity, mentioning that the role of specificity in fuzzy set theory is analogous to the role that entropy plays in probability theory. He then provides a deep analysis of the specificity of intuitionistic fuzzy sets.

9 Conclusion

Entropy and measures of information have been extensively studied for 40 years. Extensions to fuzzy sets and other representation models of uncertainty and imprecision have been proposed in many papers. These extensions are often only based on a formal similarity between the introduced quantities and classic entropies, in spite of the fact that their purpose is different, entropies measuring the decrease of uncertainty resulting from the occurrence of an event, while fuzzy set related measures evaluate the imprecision of events.

General approaches have been proposed, for instance by [13] or [4]. In this paper, we took advantage of the various quantities introduced by R. R. Yager in relation with so-called entropy, to present a means to embed them in a common approach on the basis of a view of sets and its evaluation. We introduced the notion of entropy measure and properties they can satisfy. We focused on a common general property of monotonicity which can be regarded as the most consensual common factor of most quantities introduced by R.R. Yager and others.

In the future, we propose to refine the concept of monotonicity and to use it as a key concept each time a new entropy measure must be introduced. We will take the example of evolving and dynamic systems, for which very few entropy measures have been considered and which are fundamental in the framework of machine meaning and data mining, for instance.

References

1. Aczél, J., Daróczy, Z.: On Measures of Information and their Characterizations, Mathematics in Science and Engineering, vol. 115. Academic Press, New York (1975)
2. Aczél, J., Daróczy, Z.: A mixed theory of information. I: symmetric, recursive and measurable entropies of randomized systems of events. R.A.I.R.O. Informatique théorique/Theor. Comput. Sci. 12(2), 149–155 (1978)
3. Bouchon, B.: On measure of fuzziness and information. Technical report, CNRS, Université P. et M. Curie, Paris 6 (1985). Presented at the fifth International Conference on Mathematical Modelling (Berkeley, Ca.), July 29–31 (1985)
4. Bouchon, B.: Entropic models. Cybern. Syst. Int. J. 18(1), 1–13 (1987)
5. Burillo, P., Bustince, H.: Entropy on intuitionistic fuzzy sets and on interval-valued fuzzy sets. Fuzzy Sets Syst. 78, 305–316 (1996)
6. Daróczy, Z.: Generalized information functions. Inf. Control 16, 36–51 (1970)
7. de Luca, A., Termini, S.: A definition of a nonprobabilistic entropy in the setting of fuzzy sets theory. Inf. Control 20, 301–312 (1972)
8. Guiaşu, S.: Weighted entropy. Rep. Math. Phys. 2(3), 165–179 (1971)
9. Guo, K., Song, Q.: On the entropy for Atanassov's intuitionistic fuzzy sets: an interpretation from the perspective of amount of knowledge. Appl. Soft Comput. 24, 328–340 (2014)
10. Kampé de Fériet, J.: Mesures de l'information par un ensemble d'observateurs. In: Gauthier-Villars (ed.) Comptes Rendus des Scéances de l'Académie des Sciences, série A, vol. 269, pp. 1081–1085. Paris (1969)
11. Kampé de Fériet, J.: Mesure de l'information fournie par un événement (1971). Séminaire sur les questionnaires
12. Kampé de Fériet, J.: Mesure de l'information par un ensemble d'observateurs indépendants. In: Transactions of the Sixth Prague Conference on Information Theory, Statistical Decision Functions, Random Processes - Prague, 1971, pp. 315–329 (1973)
13. Klir, G., Wierman, M.J.: Uncertainty-Based Information. Elements of Generalized Information Theory. Studies in Fuzziness and Soft Computing. Springer, Berlin (1998)
14. Mugur-Schächter, M.: The general relativity of descriptions. Analyse de Systèmes 11(4), 40–82 (1985)
15. Rényi, A.: Calcul des probabilités avec un appendice sur la théorie de l'information. Dunod (1992)
16. Szmidt, E., Kacprzyk, J.: New measures of entropy for intuitionistic fuzzy sets. In: Proceedings of the Ninth Int. Conf. on Intuitionistic Fuzzy Sets (NIFS), vol. 11, pp. 12–20. Sofia, Bulgaria (2005)
17. Yager, R.R.: Entropy measures under similarity relations. Int. J. Gen. Syst. 20(4), 341–358 (1992)
18. Yager, R.R.: Measures of entropy and fuzziness related to aggregation operators. Inf. Sci. 82(3–4), 147–166 (1995)
19. Yager, R.R.: On the entropy of fuzzy measures. IEEE Trans. Fuzzy Syst. 8(4), 453–461 (2000)

20. Yager, R.R.: Entropy and specificity in a mathematical theory of evidence. In: Yager, R., Liu, L. (eds.) Classic Works of the Dempster-Shafer Theory of Belief Functions, Studies in Fuzziness and Soft Computing, vol. 219, pp. 291–310. Springer, Berlin (2008)
21. Yager, R.R.: Some aspects of intuitionistic fuzzy sets. Fuzzy Optim. Decis. Mak. **8**, 67–90 (2009)
22. Zadeh, L.A.: The information principle. Inf. Sci. **294**, 540–549 (2015)

OWA Operators and Choquet Integrals in the Interval-Valued Setting

H. Bustince, J. Fernandez, L. De Miguel, E. Barrenechea, M. Pagola and R. Mesiar

Abstract In this chapter, we make use of the notion of admissible order between intervals to extend the definition of OWA operators and Choquet integrals to the interval-valued setting. We also present an algorithm for decision making based on these developments.

1 Introduction

Among the many contributions by Ronald Yager to Fuzzy Sets Theory, one of the most outstanding ones is that of ordered weighted aggregation operators [22] (see also [7]). These operators consider a weighting vector and carry out an ordering of the alternatives, in such a way that in fact each of the components of the weighting vector determines the relevance of each of the inputs according to its relative position with respect to all the other ones.

Applications of OWA operators have been various and successful in fields such as decision making [9, 22] or image processing [3]. However, a crucial property in

H. Bustince (✉) · J. Fernandez · L. De Miguel · E. Barrenechea · M. Pagola
Dept. Automatica Y Computacion, Universidad Publica de Navarra, Pamplona, Spain
e-mail: bustince@unavarra.es; magazine@eusflat.org

J. Fernandez
e-mail: fcojavier.fernandez@unavarra.es

L. De Miguel
e-mail: laura.demiguel@unavarra.es

E. Barrenechea
e-mail: edurne.barrenechea@unavarra.es

M. Pagola
e-mail: miguel.pagola@unavarra.es

R. Mesiar
Faculty of Civil Engineering, Department of Mathematics,
Slovak University of Technology, 81368 Bratislava, Slovakia
e-mail: mesiar@math.sk

© Springer International Publishing Switzerland 2017
J. Kacprzyk et al. (eds.), *Granular, Soft and Fuzzy Approaches for Intelligent Systems*, Studies in Fuzziness and Soft Computing 344,
DOI 10.1007/978-3-319-40314-4_4

order to build OWA operators is the fact that the usual order between real numbers is linear, and hence every possible input is comparable to every other possible input. When the corresponding algorithms are extended to other settings, such as that of interval-valued fuzzy sets, for instance, this is not true anymore, since the commonly used orders in this situation are just the partial orders.

In this chapter we present the results in [9] which allow for a consistent extension of the OWA operators to the interval-valued setting. In order to do this, we first introduce the notion of admissible order [8] and use these orders (which extend the usual partial order between intervals) for ranking appropriately the inputs.

Besides, since OWA operators are just a particular case of Choquet integrals and the construction of the latter also depends strongly on the existence of a linear order, we take advantage of our developments and recover the algorithm for decision making which was also discussed in [9]. This kind of extension is of great relevance due to the huge importance that interval-valued fuzzy sets are gaining in recent years in many applied fields [2, 5, 6, 11]. Note that approaches to this problem in the setting of averaging functions have already been considered in papers such as [4].

The structure of the chapter is as follows. In the next Section we recall some basic notions. Section 3 is devoted to the extension of OWA operators to the interval-valued setting and Sect. 4, to the corresponding developments for Choquet integrals. In Sect. 5 we recover the example in [9] to show the usefulness of the method. We finish with some concluding remarks and references.

2 Preliminaries

In this section we recall some basic concepts and notations that will be necessary for the subsequent developments. First of all we recall the notion of aggregation function in the general setting of bounded partially ordered sets.

Let $(L, \preceq)$ be a bounded partially ordered set (poset) with a smallest element (bottom) 0_L and a greatest element (top) 1_L. A mapping $A \colon L^n \to L$ is an n-ary ($n \in \mathbb{N}$, $n \geq 2$) aggregation function on $(L, \preceq)$ if it is $\preceq$-increasing, i.e., for all $\mathbf{x} = (x_1, \dots, x_n)$, $\mathbf{y} = (y_1, \dots, y_n) \in L^n$,

$$A(\mathbf{x}) \preceq A(\mathbf{y}) \text{ whenever } x_1 \preceq y_1, \dots, x_n \preceq y_n,$$

and satisfies the boundary conditions

$$A(0_L, \dots, 0_L) = 0_L, \ A(1_L, \dots, 1_L) = 1_L.$$

Remark 1 Note that for $L = [0, 1]$ and $\preceq = \leq$ the standard order of reals, we recover the usual definition of an aggregation function on the unit interval, see, e.g., [12, 15].

We denote by $L([0, 1])$ the set of all closed subintervals of the unit interval,

$$L([0, 1]) = \{[a, b] \mid 0 \leq a \leq b \leq 1\}.$$

In $L([0, 1])$ the standard partial order of intervals, i.e., the binary relation $\leq_2$, is defined by

$$[a, b] \leq_2 [c, d] \Leftrightarrow a \leq c \wedge b \leq d. \tag{1}$$

With this partial order, $(L([0, 1]), \leq_2)$ is a poset with the bottom $[0, 0]$ and top $[1, 1]$.

2.1 Admissible Orders Generated by Aggregation Functions

A crucial property for defining some types of aggregation functions on $[0, 1]$, and in particular OWA operators [22], is the possibility of comparing any two inputs. When we need to deal with intervals, however, it comes out that the order $\leq_2$ considered in the previous section is only a partial order on $L([0, 1])$. In [8], linear extensions of such orders, called admissible orders on $L([0, 1])$, were considered.

Definition 1 A binary relation $\preceq$ on $L([0, 1])$ is an *admissible order* if it is a linear order on $L([0, 1])$ refining $\leq_2$.

That is, $\preceq$ is an admissible order if for all $[a, b]$, $[c, d] \in L([0, 1])$, if $[a, b] \leq_2 [c, d]$ then also $[a, b] \preceq [c, d]$.

In [8], the generation of admissible orders on $L([0, 1])$ using appropriate pairs of aggregation functions on $[0, 1]$ was considered. Let's recall now how this was done.

Denote $K([0, 1]) = \{(a, b) \in [0, 1]^2 \mid a \leq b\}$. It is clear that intervals from $L([0, 1])$ are in a one-to-one correspondence with points from $K([0, 1])$ and a partial (linear) order $\preceq$ on one of these sets induces a partial (linear) order on the other, $[a, b] \preceq [c, d] \Leftrightarrow (a, b) \preceq (c, d)$.

Then we can define admissible orders as follows.

Proposition 1 *[8] Let $A, B: [0, 1]^2 \rightarrow [0, 1]$ be two aggregation functions, such that for all $(x, y), (u, v) \in K([0, 1])$, the equalities $A(x, y) = A(u, v)$ and $B(x, y) = B(u, v)$ can hold only if $(x, y) = (u, v)$. Define the relation $\preceq_{A,B}$ on $L([0, 1])$ by $[x, y] \preceq_{A,B} [u, v]$ if and only if*

$$A(x, y) < A(u, v)$$
$$or \ A(x, y) = A(u, v) \ and \ B(x, y) \leq B(u, v). \tag{2}$$

Then $\preceq_{A,B}$ is an admissible order on $L([0, 1])$.

A pair (A, B) of aggregation functions as in Proposition 1 which generates the admissible order $\preceq_{A,B}$ is called an admissible pair of aggregation functions. In this work we only consider admissible orders generated by continuous aggregation functions. It can be proved [8] that if (A, B) is an admissible pair of continuous aggregation functions, then there exists an admissible pair of continuous idempotent aggregation functions (A', B') such that the orders generated by the pairs (A, B) and (A', B') are the same.

Example 1 The following relations are relevant examples of admissible orders in $L([0, 1])$:

(i) $[a, b] \preceq_{Lex1} [c, d] \Leftrightarrow a < c$ or $(a = c$ and $b \le d)$,
(ii) $[a, b] \preceq_{Lex2} [c, d] \Leftrightarrow b < d$ or $(b = d$ and $a \le c)$.

The order $\preceq_{Lex1}$ is generated by the pair (P_1, P_2), where P_i, $i = 1, 2$, is the projection to the ith coordinate, and similarly, $\preceq_{Lex2}$ is generated by (P_2, P_1). These orders $\preceq_{Lex1}$ and $\preceq_{Lex2}$ are called the lexicographical orders with respect to the first or second coordinate, respectively.

A particular way of obtaining admissible orders on $L([0, 1])$, is defining them by means of the so-called Atanassov operators, K_α.

Definition 2 For $\alpha \in [0, 1]$, the operator $K_\alpha : [0, 1]^2 \to [0, 1]$ is given by

$$K_\alpha(a, b) = a + \alpha(b - a). \tag{3}$$

Note that $K_\alpha(a, b) = (1 - \alpha)a + \alpha b$, thus K_α is a weighted mean. If for $\alpha, \beta \in [0, 1]$, $\alpha \ne \beta$, we define the relation $\preceq_{\alpha, \beta}$ on $L([0, 1])$ by

$$[a, b] \preceq_{\alpha, \beta} [c, d] \Leftrightarrow K_\alpha(a, b) < K_\alpha(c, d) \text{ or } (K_\alpha(a, b) = K_\alpha(c, d) \text{ and } K_\beta(a, b) \le K_\beta(c, d)), \tag{4}$$

then it is an admissible order on $L([0, 1])$ generated by an admissible pair of aggregation functions (K_α, K_β), [8].

Proposition 2 *[8]*

(i) *Let $\alpha \in [0, 1[$. Then all admissible orders $\preceq_{\alpha, \beta}$ with $\beta > \alpha$ coincide. This admissible order will be denoted by $\preceq_{\alpha+}$.*
(ii) *Let $\alpha \in]0, 1]$. Then all admissible orders $\preceq_{\alpha, \beta}$ with $\beta < \alpha$ coincide. This admissible order will be denoted by $\preceq_{\alpha-}$.*

Remark 2

(i) The lexicographical orders $\preceq_{Lex1}$ and $\preceq_{Lex2}$ are recovered by orders $\preceq_{\alpha, \beta}$ as the orders $\preceq_{0,1} = \preceq_{0+}$ and $\preceq_{1,0} = \preceq_{1-}$, respectively.
(ii) Xu and Yager defined the order $\preceq_{XY}$ on $L([0, 1])$ by

$$[a, b] \preceq_{XY} [c, d] \Leftrightarrow a + b < c + d \ \vee \ a + b = c + d \wedge b - a \le d - c,$$

see [21]. $\preceq_{XY}$ is an admissible order which corresponds to the order $\preceq_{0.5+}$. From the statistical point of view, this order corresponds to the ordering of random variables based on the expected value as the primary criterion, and on the variance as the secondary criterion (in the case of uniform distributions this is a linear order over their supports).

3 Interval-Valued OWA Operators

As we have already discussed, we are interested in extending ordered weighted aggregation (OWA) operators introduced by Yager in [22] to the interval-valued setting. Recall that the definition of these operators strongly depends on the fact that the interval [0, 1] with the usual order between real numbers is a linearly ordered set.

Definition 3 Let $\mathbf{w} = (w_1, \dots, w_n) \in [0, 1]^n$ with $w_1 + \cdots + w_n = 1$ be a weighting vector. An ordered weighted aggregation operator $OWA_{\mathbf{w}}$ associated with $\mathbf{w}$ is a mapping $OWA_{\mathbf{w}} : [0, 1]^n \to [0, 1]$ defined by

$$OWA_{\mathbf{w}}(x_1, \dots, x_n) = \sum_{i=1}^{n} w_i x_{(i)}, \tag{5}$$

where $x_{(i)}$, $i = 1, \dots, n$, denotes the i-th greatest component of the input $(x_1, \dots, x_n)$.

However, if we make use of admissible orders, it is clear that this definition in the case of real weights can be extended straightforwardly to the interval-valued setting.

Definition 4 [9] Let $\preceq$ be an admissible order on $L([0, 1])$ and $\mathbf{w} = (w_1, \dots, w_n) \in [0, 1]^n$, $w_1 + \cdots + w_n = 1$, a weighting vector. An interval-valued OWA operator associated with $\preceq$ and $\mathbf{w}$ is a mapping $IVOWA_{\mathbf{w}}^{\preceq} : (L([0, 1]))^n \to L([0, 1])$ defined by

$$IVOWA_{\mathbf{w}}^{\preceq}([a_1, b_1], \dots, [a_n, b_n]) = \sum_{i=1}^{n} w_i \cdot [a_{(i)}, b_{(i)}], \tag{6}$$

where $[a_{(i)}, b_{(i)}]$, $i = 1, \dots, n$, denotes the i-th greatest interval of the input intervals with respect to the order $\preceq$.

Note that the arithmetic operations on intervals are given as follows:

$$w \cdot [a, b] = [wa, wb] \text{ and } [a, b] + [c, d] = [a + c, b + d].$$

Observe that IVOWA operators in Definition 4 are well defined, since

$$w_1 a_{(1)} + \dots + w_n a_{(n)} \le w_1 + \dots + w_n = 1,$$

and analogously for the upper bound. The increasing monotonicity of real-valued weighted arithmetic means ensures that the resulting set on the right-hand side of (6) is an interval $[a, b]$, $a \le b$.

Moreover, though the choice of a permutation (.) in formula (6) need not be unique (this may happen only if some inputs are repeated), the possible repetition of inputs has no influence on the resulting output interval.

First of all, we show that Definition 4 does in fact extend the usual definition of OWA operators.

Proposition 3 *[9] Let $\preceq$ be an admissible order on $L([0, 1])$ and let $\mathbf{w} = (w_1, \dots, w_n)$ $\in [0, 1]^n$ with $w_1 + \dots + w_n = 1$ be a weighting vector. Then*

$$OWA_{\mathbf{w}}(x_1, \dots, x_n) = IVOWA_{\mathbf{w}}^{\preceq}([x_1, x_1], \dots, [x_n, x_n]).$$

Remark 3 In general IVOWA operators do not preserve representability, that is, in general, the identity

$$IVOWA_{\mathbf{w}}^{\preceq}([a_1, b_1], \dots, [a_n, b_n]) = \left[OWA_{\mathbf{w}}(a_1, \dots, a_n), OWA_{\mathbf{w}}(b_1, \dots, b_n) \right], \quad (7)$$

does not hold [9].

Example 2 Consider the weighting vector $\mathbf{w} = (1, 0, 0)$ and the lexicographical order $\preceq_{Lex1}$. For the intervals $\left[\frac{1}{2}, \frac{3}{4}\right]$, $\left[\frac{1}{3}, \frac{1}{2}\right]$ and $\left[\frac{1}{3}, 1\right]$ it holds

$$\left[\frac{1}{3}, \frac{1}{2}\right] \preceq_{Lex1} \left[\frac{1}{3}, 1\right] \preceq_{Lex1} \left[\frac{1}{2}, \frac{3}{4}\right].$$

Therefore

$$IVOWA_{\mathbf{w}}^{\preceq_{Lex1}} \left(\left[\frac{1}{2}, \frac{3}{4}\right], \left[\frac{1}{3}, \frac{1}{2}\right], \left[\frac{1}{3}, 1\right] \right) = \left[\frac{1}{2}, \frac{3}{4}\right],$$

and on the other hand,

$$\left[OWA_{\mathbf{w}} \left(\frac{1}{2}, \frac{1}{3}, \frac{1}{3}\right), OWA_{\mathbf{w}} \left(\frac{3}{4}, \frac{1}{2}, 1\right) \right] = \left[\frac{1}{2}, 1\right].$$

Now, let us investigate several properties of *IVOWA* operators.

Example 3 [9] Consider the Xu and Yager order $\preceq_{XY}$ (i.e., the order $\preceq_{0.5+}$), here simply denoted by $\preceq$, and the weighting vector $\mathbf{w} = (0.8, 0.2)$. Then for intervals

$$\mathbf{x} = [0.5, 0.5], \ \mathbf{y} = [0.1, 1] \ \text{and} \ \mathbf{z} = [0.6, 0.6]$$

it holds $\mathbf{x} \preceq \mathbf{y} \preceq \mathbf{z}$ and therefore

$$IVOWA_{\mathbf{w}}^{\preceq}(\mathbf{x}, \mathbf{y}) = 0.8 \cdot [0.1, 1] + 0.2 \cdot [0.5, 0.5] = [0.18, 0.9],$$
$$IVOWA_{\mathbf{w}}^{\preceq}(\mathbf{z}, \mathbf{y}) = 0.8 \cdot [0.6, 0.6] + 0.2 \cdot [0.1, 1] = [0.5, 0.68].$$

Observe that although $\mathbf{x} = [0.5, 0.5] \leq_2 [0.6, 0.6] = \mathbf{z}$ (i.e., we have increased the first input interval with respect to the order $\leq_2$), the obtained values of the *IVOWA*$_{\mathbf{w}}^{\preceq}$ operator are not comparable in the order $\leq_2$, i.e., *IVOWA*$_{\mathbf{w}}^{\preceq}$ is not an aggregation function with respect to $\leq_2$.

The following examples and results are taken from [9].

Example 4 Consider the order $\preceq_{A,B}$ generated by an admissible pair (A, B) of aggregation functions, where $A(x, y) = (\sqrt{x} + \sqrt{y})/2$ and $B(x, y) = y$, and the IVOWA operator associated with the weighting vector $\mathbf{w} = \left(\frac{2}{3}, \frac{1}{3}\right)$. Let

$$\mathbf{x} = [0.25, 0.25], \quad \mathbf{y} = [0, 1], \quad \mathbf{z} = [0.25, 0.28].$$

Then $\mathbf{x} \preceq_{A,B} \mathbf{y} \preceq_{A,B} \mathbf{z}$ and

$$IVOWA_{\mathbf{w}}^{\preceq_{A,B}}(\mathbf{x}, \mathbf{y}) = \frac{2}{3}\mathbf{y} + \frac{1}{3}\mathbf{x} = \left[\frac{1}{12}, \frac{3}{4}\right],$$

$$IVOWA_{\mathbf{w}}^{\preceq_{A,B}}(\mathbf{z}, \mathbf{y}) = \frac{2}{3}\mathbf{z} + \frac{1}{3}\mathbf{y} = \left[\frac{1}{6}, 0.52\right].$$

Next, $A\left(\frac{1}{12}, \frac{3}{4}\right) = 0.57735$ and $A\left(\frac{1}{6}, 0.52\right) = 0.5646679$, which means that $IVOWA_{\mathbf{w}}^{\preceq_{A,B}}(\mathbf{x}, \mathbf{y}) >_{A,B} IVOWA_{\mathbf{w}}^{\preceq_{A,B}}(\mathbf{z}, \mathbf{y})$ and this contradicts the $\preceq_{A,B}$-increasing monotonicity of $IVOWA_{\mathbf{w}}^{\preceq_{A,B}}$ operator.

Let us write now $K_\alpha([a, b])$ to mean that we have assigned to an interval $[a, b] \in L([0, 1])$ the same value as to the corresponding point $(a, b) \in K([0, 1])$ by the mapping K_α, i.e., $K_\alpha([a, b]) = a + \alpha(b - a)$.

Proposition 4 *Let $\preceq$ be an admissible order on $L([0, 1])$ generated by a pair (K_α, B) and let $IVOWA_{\mathbf{w}}^{\preceq}$ be an interval-valued OWA operator defined by (6). Then*

$$K_\alpha\left(IVOWA_{\mathbf{w}}^{\preceq}([a_1, b_1], \ldots, [a_n, b_n])\right) = OWA_{\mathbf{w}}\left(K_\alpha([a_1, b_1]), \ldots, K_\alpha([a_n, b_n]),\right) \quad (8)$$

independently of B.

Corollary 1 *Let $\preceq_{\alpha,\beta}$ be an admissible order on $L([0, 1])$ introduced in (4). Then the interval-valued OWA operator $IVOWA_{\mathbf{w}}^{\preceq_{\alpha,\beta}}$ is an aggregation function on $L([0, 1])$ with respect to the order $\preceq_{\alpha,\beta}$.*

Due to the relation between interval-valued fuzzy sets and Atanassov intuitionistic fuzzy sets, IVOWA operators can be seen as modified and special cases of intuitionistic OWA operators, see, e.g. [17, 23]. However, the approaches in both mentioned papers are different from the one presented here, as the aggregation of intervals is split into the aggregation of their left bounds (membership functions of intuitionistic fuzzy sets) and aggregation of right bounds (complements to non-membership functions).

Besides, it is worth to mention that OWA operators have been recently extended to complete lattices in [16]. As a particular case, OWA operators on intervals in the form (7) are obtained.

4 Interval-Valued Choquet Integral

4.1 Interval-Valued Choquet Integral Based on Aumann's Approach

Recall that OWA operators are a particular case of the so-called Choquet integrals. Like the former, the latter can also be extended to the interval-valued case by means of admissible orders. In this section we propose such extension for discrete interval-valued Choquet integrals of interval-valued fuzzy sets based on admissible orders $\preceq_{A,B}$. In the first subsection, we recall an extension of the Choquet integral to the interval-valued setting, which has been discussed, e.g., in [13, 26]. A similar idea led Aumann [1] to introducing his integral of set-valued functions. These concepts are of the same nature as is the Zadeh extension principle [25].

Definition 5 Let $U \neq \emptyset$ be a finite set. A fuzzy measure m is a set function $m\colon 2^U \to [0, 1]$ such that

$$m(\emptyset) = 0, \quad m(U) = 1, \text{ and } m(A) \le m(B) \text{ whenever } A \subseteq B.$$

Definition 6 The discrete Choquet integral (or expectation) of a fuzzy set $f\colon U \to [0, 1]$ with respect to a fuzzy measure m is defined by

$$C_m(f) = \sum_{i=1}^{n} f(u_{\sigma(i)}) \left(m\left(\{u_{\sigma(i)}, \dots, u_{\sigma(n)}\} \right) - m\left(\{u_{\sigma(i+1)}, \dots, u_{\sigma(n)}\} \right) \right),$$

where $\sigma\colon \{1, \dots, n\} \to \{1, \dots, n\}$ is a permutation such that

$$f(u_{\sigma(1)}) \le f(u_{\sigma(2)}) \le \cdots \le f(u_{\sigma(n)}),$$

and $\{u_{\sigma(n+1)}, u_{\sigma(n)}\} = \emptyset$, by convention.

Now we can make the extension of this definition to the interval-valued setting as follows.

Definition 7 Let $F\colon U \to L([0, 1])$ be an interval-valued fuzzy set and $m\colon 2^U \to [0, 1]$ a fuzzy measure. The discrete Choquet integral $\mathbf{C}_m(F)$ of an interval-valued fuzzy set F with respect to m is given by

$$\mathbf{C}_m(F) = \{C_m(f) \mid f\colon U \to [0, 1], f(u_i) \in F(u_i)\}. \tag{9}$$

Remark 4 From the properties of the standard Choquet integral of fuzzy sets it follows that

$$\mathbf{C}_m(F) = \left[C_m(f_*), C_m(f^*) \right], \tag{10}$$

where $f_*, f^*\colon U \to [0, 1]$ are given by $f_*(u_i) = a_i$ and $f^*(u_i) = b_i$, and $[a_i, b_i] = F(u_i)$.

Several properties of the discrete interval-valued Choquet integral $\mathbf{C}_m$ are discussed in [13, 26]. For example, this integral is comonotone additive, i.e.,

$$\mathbf{C}_m(F + G) = \mathbf{C}_m(F) + \mathbf{C}_m(G),$$

whenever $F, G: U \to L([0, 1])$ are such that interval $F(u_i) + G(u_i) \subseteq [0, 1]$ for each $u_i \in U$, and F, G are comonotone, i.e.,

$$(f^*(u_i) - f^*(u_j))(g^*(u_i) - g^*(u_j)) \geq 0$$

and

$$(f_*(u_i) - f_*(u_j))(g_*(u_i) - g_*(u_j)) \geq 0$$

for all $u_i, u_j \in U$.

4.2 Interval-Valued Choquet Integral with Respect to $\preceq_{A,B}$-orders

The basic idea of the original Choquet integral [10] is based on the linear order of reals allowing two different looks at functions. The vertical look is based on function values and it is a background of the Lebesgue integral, while the horizontal look is linked to level cuts and it is a basis not only for the Choquet integral, but also for several other types of integrals, see [14], including among others, the Sugeno integral [19]. In this subsection we introduce a discrete interval-valued Choquet integral of interval-valued fuzzy sets based on an (admissible) order of intervals in $L([0, 1])$ directly, without using the notion of the Choquet integral of scalar-valued fuzzy sets.

Let $\preceq_{A,B}$ be an admissible order on $L([0, 1])$ given by a generating pair of aggregation function (A, B) as explained in Proposition 1. The discrete interval-valued Choquet with respect to the order $\preceq_{A,B}$ is defined as follows.

Definition 8 [9] Let $F: U \to L([0, 1])$ be an interval-valued fuzzy set and $m: 2^U \to [0, 1]$ a fuzzy measure. The discrete interval-valued Choquet integral with respect to an admissible order $\preceq_{A,B}$ ($\preceq_{A,B}$-Choquet integral for short) of an interval-valued fuzzy set F with respect to m, with the notation $\mathbf{C}_m^{\preceq_{A,B}}(F)$, is given by

$$\mathbf{C}_m^{\preceq_{A,B}}(F) = \sum_{i=1}^{n} F(u_{\sigma_{A,B}(i)}) \left(m\left(\{u_{\sigma_{A,B}(i)}, \ldots, u_{\sigma_{A,B}(n)}\} \right) \right) - \left(m\left(\{u_{\sigma_{A,B}(i+1)}, \ldots, u_{\sigma_{A,B}(n)}\} \right) \right) \quad (11)$$

where $\sigma_{A,B}: \{1,\dots,n\} \to \{1,\dots,n\}$ is a permutation such that

$$F(u_{\sigma_{A,B}(1)}) \leq F(u_{\sigma_{A,B}(2)}) \leq \cdots \leq F(u_{\sigma_{A,B}(n)}),$$

and $\{u_{\sigma_{A,B}(n+1)}, u_{\sigma_{A,B}(n)}\} = \emptyset$, by convention.

Remark 5 Observe that if $F(u_i) = [a_i, b_i]$, $i = 1,\dots,n$, then (11) can be written as

$$\mathbf{C}_m^{\preceq_{A,B}}(F) = \left[\sum_{i=1}^{n} a_{\sigma_{A,B}(i)} \left(m\left(\left\{u_{\sigma_{A,B}(i)}, \dots, u_{\sigma_{A,B}(n)}\right\}\right) - m\left(\left\{u_{\sigma_{A,B}(i+1)} - u_{\sigma_{A,B}(n)}\right\}\right)\right), \right.$$

$$\left. \sum_{i=1}^{n} b_{\sigma_{A,B}(i)} \left(m\left(\left\{u_{\sigma_{A,B}(i)}, \dots, u_{\sigma_{A,B}(n)}\right\}\right) - m\left(\left\{u_{\sigma_{A,B}(i+1)} - u_{\sigma_{A,B}(n)}\right\}\right)\right) \right].$$

Next, for any fixed $F: U \to L([0,1])$ such that the corresponding f_* and f^* are comonotone, i.e., for all $u_i, u_j \in U$,

$$\left(f_*(u_i) - f_*(u_j)\right)\left(f^*(u_i) - f^*(u_j)\right) \geq 0,$$

it holds that for any admissible pair (A,B) of aggregation functions the Choquet integrals of F introduced in Definitions 7 and 8 coincide, i.e., $\mathbf{C}_m^{\preceq_{A,B}}(F) = \mathbf{C}_m(F)$.

Remark 6 The concept of an interval-valued Choquet integral $\mathbf{C}_m^{\preceq_{A,B}}$ introduced in Definition 8 extends the standard discrete Choquet integral given by (6). Indeed, if $F: U \to L([0,1])$ is singleton-valued, i.e., it is a fuzzy subset of U, then

$$\mathbf{C}_m^{\preceq_{A,B}}(F) = \mathbf{C}_m(F) = C_m(F)$$

independently of A and B.

Moreover, observe that if m is a symmetric fuzzy measure [20] then, similarly to the classical case, $\mathbf{C}_m^{\preceq_{A,B}} = IVOWA_m^{\preceq_{A,B}}$, where $\mathbf{w} = (w_1, \dots, w_n)$, $w_{n-i+1} = m(\{i, i+1, \cdots, n\}) - m(\{i+1, \cdots, n\})$, $i = 1, \cdots, n$, with convention $\{n+1, n\} = \emptyset$.

5 Application to Multi-expert Decision Making

In this section we recover the method explained in [9]. Consider n experts $E = \{e_1, \dots, e_n\}$, $(n > 2)$ and a set of p alternatives $X = \{x_1, \dots, x_p\}$, $(p \geq 2)$. Our goal is to find the alternative which is the most accepted one by the n experts.

Many times experts have difficulties to determine the exact value of the preference of alternative x_i against x_j for each $i, j \in \{1, \dots, p\}$. When this happens, they usually give their preferences by means of elements in $L([0,1])$; that is, by means of intervals. In these cases we say that the preference of the experts is given by a numerical value inside the interval.

5.1 Interval-Valued Preference Relations

An interval-valued fuzzy binary relation R_{IV} on X is an interval-valued fuzzy subset of $X \times X$; that is, $R_{IV} : X \times X \rightarrow L([0,1])$. The interval $R_{IV}(x_i, x_j) = R_{IV_{ij}}$ denotes the degree to which elements x_i and x_j are related in the relation R_{IV} for all $x_i, x_j \in X$. Particularly, in preference analysis, $R_{IV_{ij}}$ denotes the degree to which alternative x_i is preferred to alternative x_j.

Each expert e provides his/her preferences by means of an interval-valued fuzzy relation R_{IV_e} with p rows and p-columns and where the elements in the diagonal are not considered; that is,

$$
R_{IV_e} = \begin{pmatrix}
 & x_1 & x_2 & \cdots, & \cdots, & x_p \\
x_1 & - & [\underline{R}_{e_{12}}, \overline{R}_{e_{12}}] & [\underline{R}_{e_{13}}, \overline{R}_{e_{13}}] & \cdots, & [\underline{R}_{e_{1p}}, \overline{R}_{e_{1p}}] \\
x_2 & [\underline{R}_{e_{21}}, \overline{R}_{e_{21}}] & - & [\underline{R}_{e_{23}}, \overline{R}_{e_{23}}] & \cdots, & [\underline{R}_{e_{2p}}, \overline{R}_{e_{2p}}] \\
 & \cdots & \cdots & \cdots & & \cdots \\
x_p & [\underline{R}_{e_{p1}}, \overline{R}_{e_{p1}}] & [\underline{R}_{e_{p2}}, \overline{R}_{i_{p2}}] & \cdots & \cdots & -
\end{pmatrix}
$$

Then we can consider the following algorithm (Algorithm 1).

1. Choose a linear order $\preceq$ between intervals.
2. Choose a weighting vector $\mathbf{w}$.
3. Calculate the interval-valued collective fuzzy relation R_{IV_C} using the operators $IVOWA_{\mathbf{w}}^{\preceq}$.
4. For each row i in R_{IV_C} build the fuzzy measure m_i:

$$
m_i(\{x_{ij}\})_{i \neq j} = \left(\frac{\underline{R}_{ij} + \overline{R}_{ij}}{\sum_{\substack{l=1 \\ l \neq i}}^{p} (\underline{R}_{il} + \overline{R}_{il})} \right)^2
$$

$$
m_i(\{x_{ij}, x_{ik}\})_{\substack{i \neq j \\ i \neq k \\ j < k}} = \left(\frac{\underline{R}_{ij} + \overline{R}_{ij} + \underline{R}_{ik} + \overline{R}_{ik}}{\sum_{\substack{l=1 \\ l \neq i}}^{p} (\underline{R}_{il} + \overline{R}_{il})} \right)^2 \tag{12}
$$

$$\cdots$$

that is, given $i \in \{1, \dots, p\}$, for every $A \subseteq \{1, \dots, n\} \setminus \{i\}$

$$
m_i(\{x_{ij} \mid j \in A\}) = \left(\frac{\sum_{j \in A} \underline{R}_{ij} + \overline{R}_{ij}}{\sum_{\substack{l=1 \\ l \neq i}}^{p} (\underline{R}_{il} + \overline{R}_{il})} \right)^2
$$

5. For each row of R_{IV_C} aggregate the intervals by means of the interval-valued Choquet integral constructed with the order $\preceq$ chosen in step $IVD1$) and the measure built in step $IVD4$).

6. Take as solution the alternative corresponding to the row with the biggest interval with respect to the order $\leq$ chosen in step *IVD*1).

Note that steps (1)–(3) correspond to the aggregation phase whereas the other steps correspond to the exploitation phase.

Remark 7 If the preference relations provided by the experts are numerical, then the proposed algorithm recovers the classical methods for multi-expert decision making which make use of Choquet integrals in the exploitation phase [24].

Proposition 5 *The measure defined in Eq. (12) is superadditive; that is, for any two nonintersecting subsets $A, B \in X$, $A \cap B = \emptyset$,*

$$m_i(A \cup B) \geq m_i(A) + m_i(B) \text{ for each row } i = 1, \cdots, p \tag{13}$$

Proof The fact follows from the superadditivity of the quadratic function $f(x) = x^2$ on $[0, 1]$. □

Remark 8 Note that we do not require that R_{IV_e} is reciprocally additive; that is we do not demand the following property:

$$\underline{R}_{e_{ij}} + \overline{R}_{e_{ji}} = 1 \text{ and } \underline{R}_{e_{ji}} + \overline{R}_{e_{ij}} = 1$$

The advantage of not demanding this property is that we do not modify the preferences provided by the experts in order to ensure additivity.

The result of Algorithm 1 depends of the order $\leq$ and the weighting vector **w** that we use. In both cases, the choice we make is linked to the application in which we are working. Usually, the choice of the weighting vector is easier, since the weights are often related to the quantifiers given in [22] and it is the application which determines that we have to consider aggregations of the type : *most of the experts* say ... or *at least one half of the experts* say ... etc.

The choice of the order is more complicate. Both the application and the experts should be taken into account. For instance, if the experts are considered to be optimistic, it may be logical to use the order $\leq_{Lex2}$. On the contrary, if they are considered to be pessimistic, the order $\leq_{Lex1}$ might be more suitable. However, in many cases we do not have this information. Clearly, if the application determines the order to be used, we apply Algorithm 1 directly.

If we do not know which is the most appropriate order, we propose to run Algorithm 1 with different orders; for instance, with s different orders. If for all the considered orders we obtain the same result, that is, the same alternative, then we have finished and we choose as the winning alternative that one. However, if we obtain different winning alternatives, then we propose the following algorithm (Algorithm 2):

1. Run Algorithm 1 for each of the s selected orders;
2. For each interval-valued collective fuzzy relation $R_{IV_c}^{\ l}$ with $l = 1, \ldots, s$, calculate the fuzzy preference relation such that each of its elements is obtained as the midpoint of the corresponding interval in the relation $R_{IV_c}^{\ l}$;
3. Calculate the arithmetic mean matrix MP of the s fuzzy matrices obtained in Step SC2.

$$MP = \begin{pmatrix} - & a_{12} & a_{13} & \cdots, & a_{1p} \\ a_{21} & - & a_{23} & \cdots, & a_{2p} \\ \cdots & \cdots & \cdots & - & \cdots \\ a_{p1} & a_{p2} & \cdots & a_{p(p-1)} & - \end{pmatrix}$$

4. Build the measure:

$$m(\{x_i\}) = \left(\frac{a_{i1}+\cdots+a_{i(i-1)}+a_{i(i+1)}+\cdots a_{ip}}{a_{12}+\cdots+a_{1p}+\cdots+a_{i1}+\cdots+a_{ip}+\cdots+a_{p1}+\cdots+a_{i(p-1)}} \right)^2$$

$$m(\{x_i, x_j\}) = \left(\frac{a_{i1}+\cdots+a_{i(i-1)}+a_{i(i+1)}+\cdots+a_{ip}+a_{j1}+\cdots+a_{j(j-1)}+a_{j(j+1)}+\cdots a_{jp}}{a_{12}+\cdots+a_{1p}+\cdots+a_{i1}+\cdots+a_{ip}+\cdots+a_{p1}+\cdots+a_{i(p-1)}} \right)^2 \quad (14)$$

$\cdots$

that is, for each $A \subseteq \{1, \ldots, p\}$

$$m(\{x_i \mid i \in A\}) = \left(\frac{\sum_{i \in A} \sum_{j \in \{1,\ldots,p\} \setminus \{i\}} a_{ij}}{\sum_{i=1}^n \sum_{j \in \{1,\ldots,p\} \setminus \{i\}} a_{ij}} \right)^2$$

5. Using the measure m from step SC4 calculate the Shapley value:

$$\varphi(x_i) = \sum_{A \subseteq X \setminus \{x_i\}} \frac{1}{n \binom{n-1}{|A|}} (m(A \cup \{x_i\}) - m(A)) \quad (15)$$

for each of the solutions obtained in step SC1.
6. Take as solution the alternative corresponding to the highest Shapley value.

Remark 9

(i) Once the winning alternatives x_i have been calculated with Algorithm 1 ($i = 1, \cdots, s$), the Shapley value $\varphi(x_i)$ measures the relevance of alternative x_i in possible coalitions with other alternatives.

(ii) The advantage of using the measure given in Eq. (14) is that it takes into account all the preference values provided by all the experts. In this way, the Shapley value is calculated using the same matrix MP for all the winning alternatives. This is the main difference between the measure given in Eq. (12) and the one given in Eq. (14). Note that the measure in Eq. (14) is superadditive too.

It may happen that with Algorithm 2 we get the same Shapley value for different alternatives and we can not decide which is the best one. Then we can take as the solution the one which appears most times as winner when we run Algorithm 1 with the s different orders.

6 Concluding Remarks

In this work, we have reviewed the results in [9]. In particular, by means of the notion of admissible order, we have extended OWA operators and Choquet integrals to the interval-valued setting. The interest in this definition lies in the fact that admissible orders enable us to build many different OWA operators, that, on one hand, extend usual operators, but, on the other hand, leave some free space to choose the most appropriate one for the problem under consideration. The question of determining the most suitable linear order for a given problem is of great interest, as we have exhibited for multi-expert decision making when we use intervals to represent the alternatives.

Note that our approach can be further generalized. Indeed, instead of admissible orders generated by a couple (A, B) of aggregation functions one can deal with a couple of two weak orders $(\leq^*, \leq^{**})$ compatible with the standard partial order $\leq_2$ (recall that a weak order $\leq$ on $L([0, 1])$ is a reflexive and transitive relation such that for any two intervals $[a, b]$ and $[c, d]$ from $L([0, 1])$, either $[a, b] \leq [c, d]$ or $[c, d] \leq [a, b]$; moreover, $\leq$ is compatible with $\leq_2$ if from the relation $[a, b] \leq_2 [c, d]$ follows that $[a, b] \leq [c, d]$), such that $[a, b] =^* [c, d]$ and $[a, b] =^{**} [c, d]$ only if $[a, b] = [c, d]$. Clearly, our (A, B) approach is a particular case of the proposed generalization, when considering $[a, b] \leq^* [c, d]$ if and only if $A(a, b) \leq A(c, d)$, and $[a, b] \leq^{**} [c, d]$ if and only if $B(a, b) \leq B(c, d)$.

Acknowledgments The first two authors were supported by the project TIN2013-40765-P of the Spanish Ministry of Science. A. Kolesárová and R. Mesiar were supported by grant APVV-14-0013.

References

1. Aumann, R.J.: Integrals of set-valued functions. J. Math. Anal. Appl. **12**, 1–12 (1965)
2. Barrenechea, E., Bustince, H., De Baets, B., Lopez-Molina, C.: Construction of interval-valued fuzzy relations with application to the generation of fuzzy edge images. IEEE Trans. Fuzzy Syst. **19**(5), 819–830 (2011)
3. Beliakov, G., Bustince, H., Paternain, D.: Image reduction using means on discrete product lattices. IEEE Trans. Image Process. **21**(3), 1070–1083 (2012)
4. Beliakov, G., Bustince, H., Goswami, D.P., Mukherjee, U.K., Pal, N.R.: On averaging operators for Atanassov's intuitionistic fuzzy sets. Inf. Sci. **181**, 1116–1124 (2010)
5. Bustince, H.: Interval-valued fuzzy sets in soft computing. Int. J. Comput. Intell. Syst. **3**(2), 215–222 (2010)

6. Bustince, H., Barrenechea, E., Pagola, M., Fernandez, J.: Interval-valued fuzzy sets constructed from matrices: application to edge detection. Fuzzy Sets Syst. **160**, 1819–1840 (2009)
7. Bustince, H., Calvo, T., De Baets, B., Fodor, J., Mesiar, R., Montero, J., Paternain, D., Pradera, A.: A class of aggregation functions encompassing two-dimensional OWA operators. Inf. Sci. **180**, 1977–1989 (2010)
8. Bustince, H., Fernandez, J., Kolesárová, A., Mesiar, R.: Generation of linear orders for intervals by means of aggregation functions. Fuzzy Sets Syst. **220**, 69–77 (2013)
9. Bustince, H., Galar, M., Bedregal, B., Kolesárová, A., Mesiar, R.: A new approach to interval-valued Choquet integrals and the problem of ordering in interval-valued fuzzy sets applications. IEEE Trans. Fuzzy Syst. **21**, 1150–1162 (2013)
10. Choquet, G.: Theory of capacities. Annales de l'Institute Fourier **5**, 131–292 (1953–54)
11. Galar, M., Fernandez, J., Beliakov, G., Bustince, H.: Interval-valued fuzzy sets applied to stereo matching of color images. IEEE Trans. Image Process. **20**, 1949–1961 (2011)
12. Grabisch, M., Marichal, J.-L., Mesiar, R., Pap, E.: Aggregation Functions. Cambridge University Press, Cambridge (2009)
13. Jang, L.C.: Interval-valued Choquet integrals and their applications. J. Appl. Math. Comput. **16**, 429–443 (2004)
14. Klement, E.P., Mesiar, R., Pap, E.: A universal integral as common frame for Choquet and Sugeno integral. IEEE Trans. Fuzzy Syst. **18**(1), 178–187 (2010)
15. Komorníková, M., Mesiar, R.: Aggregation functions on bounded partially ordered sets and their classification. Fuzzy Sets Syst. **175**, 48–56 (2011)
16. Lizasoain, I., Moreno, C.: OWA operators defined on complete lattices. Fuzzy Sets Syst. **224**, 36–52 (2013)
17. Mitchel, H.B.: An intuitionistic OWA operator. Int. J. Uncertainty, Fuzziness Knowl. Based Syst. **12**, 843–860 (2004)
18. Shapley, L.S.: A Value for n-Person Game. Princeton University Press, Princeton (1953)
19. Sugeno, M.: Theory of Fuzzy Integrals and Its Applications. Ph.D. thesis, Tokyo Institute of Technology (1974)
20. Wang, Z., Klir, G.J.: Fuzzy Measure Theory. Plenum Press, New York (1992)
21. Xu, Z.S., Yager, R.: Some geometric aggregation operators based on intuitionistic fuzzy sets. Int. J. Gen. Syst. **35**, 417–433 (2006)
22. Yager, R.: On ordered weighted averaging aggregation operators in multi-criteria decision making. IEEE Trans. Syst. Man Cybern. **18**, 183–190 (1988)
23. Yager, R.: OWA aggregation of intuitionistic fuzzy sets. Int. J. Gen. Syst. **38**, 617–641 (2009)
24. Yager, R.: On prioritized multiple-criteria aggregation. IEEE Trans. Syst. Man Cybern. B Cybern. **42**(5), 1297–1305 (2012)
25. Zadeh, L.A.: The concept of a linguistic variable and its applications to approximate reasoning. Inf. Sci. **8**, Part I, 199–251 (1975), Part II, 301–357, **9**, Part III, 43–80
26. Zhang, D., Wang, Z.: On set-valued fuzzy integrals. Fuzzy Sets Syst. **56**, 237–247 (1993)

Information Theory Applications in Soft Computing

Paul Elmore and Frederick Petry

Abstract An overview of information theory metrics and the ranges of their values for extreme probability cases is provided. Imprecise database models including similarity based fuzzy models and rough set models are described. Various entropy measures for these database models' content and responses to querying is provided. Aggregation of uncertainty representations are also considered. In particular the possibilistic conditioning of probability aggregation is examined. Information measures are used to compare the resultant conditioned probability to the original probability for three cases of possibility distributions.

Keywords Shannon entropy · Gini index · Fuzzy database · Rough database · Possibility distribution · Possibilistic conditioning

1 Introduction

Information theory metrics have been utilized for various topics ranging from biology to computer science. In communication theory, Shannon [1] introduced the entropy concept which was used to characterize signal information content. Since then, variations of these information theoretic measures have been successfully applied to applications in many diverse fields. In particular, the representation of imprecision with entropy metrics has been applied to all areas of databases, including fuzzy database querying [2], data allocation [3], classification in rule-based systems[4], and measuring uncertainty in rough and fuzzy rough relational databases [5].

P. Elmore · F. Petry (✉)
Naval Research Laboratory, Stennis Space Center, Gulfport, USA
e-mail: fpetry@nrlssc.navy.mil

© Springer International Publishing Switzerland 2017
J. Kacprzyk et al. (eds.), *Granular, Soft and Fuzzy Approaches for Intelligent Systems*, Studies in Fuzziness and Soft Computing 344, DOI 10.1007/978-3-319-40314-4_5

81

In this chapter we provide in Sect. 2 definitions of Shannon entropy, Gini index and Renyi entropy. The ranges of possible values for these are developed. Next uncertain database models including those based on fuzzy similarity relationships and rough sets are described. Information measures for these models are then provided. The last section describes the use of information measure to evaluate the results of an aggregation approach, possibilistic conditioning of probability.

2 Information Theory Background

The term "information" has a number of various meanings, mostly imprecise, depending on the exact context in which it is used. One very general definition is that information is a measure of the value of data as used in decision making [6]. In this section we will review information metrics that can be used to evaluate uncertainty in databases and aggregation. Shannon's entropy has been a commonly accepted standard for information measures; however, the concept of information is so rich and broad that multiple approaches to the quantification of information are desirable [7, 8]. Thus in this section, we will also examine other measures, such as the Gini index and Renyi entropy.

2.1 Shannon Entropy

Shannon entropy has been the most commonly applied measure of randomness for information content [1]. For a probability distribution $P = \{p_1, p_2, \ldots p_n\}$ this is given as

$$S(P) = - \sum_{i=1}^{n} p_i ln(p_i)$$

The well-known minimum and maximum values for the Shannon entropy are considered for the extreme probability cases of complete certainty and complete uncertainty.

First, for complete certainty, P_{cc}, we have for some t, $p_t = 1$, and so

$$S(P_{cc}) = - (1\, ln\,(1) + \sum_{i=1,\, \neq t}^{n} 0\, ln\,(0)) = 1 * 0 + \sum_{i=1,\, \neq t}^{n} 0 = 0$$

Note this follows as $\lim_{p \to 0+} p\, ln\, p = 0$. So if a probability distribution represents complete certainty, there is no uncertainty, i.e. maximum information.

Next, for the case of complete uncertainty represented by an equi-probable distribution, P_{cu}, where $\forall i$, $p_i = 1/n$.

$$S(P_{cu}) = -\sum_{i=1}^{n} \frac{1}{n} * ln\left(\frac{1}{n}\right) = -\frac{1}{n}\sum_{i=1}^{n}(ln(1) - ln(n)) = -n*\frac{1}{n}(0 - ln(n)) = ln(n).$$

That is, when all probabilities are equi-probable, this is the most unpredictable, uncertain situation, and so represents the minimum information. In summary, the range of Shannon's entropy for a given probability distribution P is:

$$0 \le S(P) \le ln(n).$$

2.2 Gini Index

The Gini index, $G(P)$, also known as the Gini coefficient, is a measure of statistical dispersion developed by Gini [9], and it can be given as

$$G(P) = 1 - \sum_{i=1}^{n} p_i^2.$$

Some practitioners use $G(P)$ versus $S(P)$ since it does not involve a logarithm, making analytic solutions simpler. The Gini index has been used in consideration of inequalities in various areas such as economics, ecology and engineering [10]. A very important application of the Gini index is as a splitting criterion for the decision tree induction in machine learning and data mining [11].

It is accepted in practice for diagnostic test selection that the Shannon and Gini measures are interchangeable [12]. The specific relationship of Shannon entropy and the Gini index has been discussed in the literature [13]. Theoretical support for this practice is provided in Yager's independent consideration of alternative measures of entropy [14] where he derives the same form for an entropy measure as the Gini measure.

Now as it is done for Shannon entropy, we consider the maximum and minimum values for $G(P)$. Letting $R = \sum_{i=1}^{n} p_i^2$ then since $0 \le p_i \le 1$ $(0 \le p_i^2 \le 1)$ and at least one $p_i > 0$, $0 < R \le 1$. $R = 1$ only if for some t, $p_t = 1$. Thus, $G(P) > 0$ unless $p_t = 1$ where $G(P) = 0$. This is the case for the distribution P_{cc}, since $p_t = 1$, $p_i = 0$, $i \neq t$. Specifically

$$G(P_{cc}) = 1 - (p_t^2 + \sum_{i \neq 1}^{n} p_i^2) = 1 - (1^2 + 0) = 0$$

As for the Shannon entropy this corresponds to no uncertainty and has the same value of 0.

Next we examine the index for the equi-probable distribution, P_{cu}, where $p_i = 1/n$ for all i.

$$G(P_{cu}) = 1 - \sum_{i=1}^{n} (1/n)^2 = 1 - n(1/n^2) = 1 - 1/n = (n-1)/n$$

As n increases, $n \to \infty$, $G(P_{cu}) \to 1$. Thus, in the case of an equiprobable distribution, we have increasing values for $G(P_{cu})$ with n, and in general the range for $G(P)$ is

$$0 \leq G(P) \leq (n-1)/n < 1.$$

2.3 Renyi Entropy

Renyi introduced a parameterized family of entropies as a generalization of Shannon entropy [15, 16] The intention was to have the most general approach that preserved the additivity property and satisfied the probability axioms of Kolmogorov. Renyi entropy is

$$S_\alpha(P) = \frac{1}{1-a} * ln(\sum_{i=1}^{n} p_i^\alpha).$$

Cases of the parameter α:

$\alpha = 0: S(P) = ln|P|$—Hartley Entropy [17]

$lim\, \alpha \to 1: S_1(P) = - \sum_{i=1}^{n} p_i * ln(p_i)$—Shannon Entropy

$\alpha = 2: S_2(P) = - ln\left(\sum_{i=1}^{n} p_i^2 \right)$—Collision, or quadratic entropy

$\alpha \to \infty: S_\infty(P) = \underset{i=1}{\overset{n}{Min}}(ln\, p_i) = - \underset{i=1}{\overset{n}{Max}}(ln\, p_i).$

This last case is the smallest entropy in the Renyi family and so is the strongest way to obtain an information content measure. It is never larger than the Shannon entropy. Thus, the possible ranges of α capture the following:

High α: high probability events,
Low α: weight possible events more equally,
$\alpha = 0, 1 \to$ Hartley or Shannon, respectively.

3 Information-Theoretic Measures for Fuzzy and Rough Database Models

The fuzzy and rough database models both have several features in common with ordinary relational databases. Both represent data as a collection of relations containing tuples. These relations are sets. The tuples of a relation are its elements, and like elements of sets in general, are unordered and nonduplicated. A tuple t_i takes the form $(d_{i1}, d_{i2}, ..., d_{im})$, where d_{ij} is a domain or attribute value of a particular domain set D_j. In the ordinary relational database, $d_{ij} \in D_j$. In the fuzzy and rough database, however, as in other non-first normal form extensions to the relational model [18, 19] $d_{ij} \subseteq D_j$, and although it is not required that d_{ij} be a singleton, $d_{ij} \neq \emptyset$.

3.1 Fuzzy Databases

Fuzzy databases are found in areas which involve some imprecision or uncertainty in the data and in decision-making utilization of the data [20]. In order to help understand the impact of such imprecision, information-theoretic characterizations have been developed which measure the overall uncertainty in an entire relation. Additionally, a variation of fuzzy entropy has been used to determine how well a fuzzy query differentiates among potential responses.

3.2 Fuzzy Similarity Model

The use of similarity relationships in a relational model was developed by Buckles and Petry [21]. This approach attempts to generalize the concept of null and multiple-valued domains for implementation within an environment consistent with the relational algebra. In fact, the nonfuzzy relational database is a special case of this fuzzy relational database approach.

A similarity relation, $s(x, y)$, for given domain, D, is a mapping of every pair of elements in the domain onto the unit interval:

$$s_D(x, y): D \to [0, 1]$$

There are three basic properties for a similarity relation, $x, y, z \in D$ [22]:

1. Reflexive: $s_D(x, x) = 1$
2. Symmetric: $s_D(x, y) = s_D(y, x)$
3. Transitive: $s_D(x, z) \geq Max(Min[s_D(x, y), s_D(y, z)]) : (T1)$

This particular max-min form of transitivity is known as $T1$ transitivity. Another useful form is $T2$ also known as max-product:

3'. Transitive: $s_D(x, z) = Max([s_D(x, y) * s_D(y, z)]) : (T2)$

where $*$ is arithmetic multiplication.

An example of a similarity relation satisfying $T2$ transitivity, where $\beta > 0$ is an arbitrary constant and $x, y \in D$, is:

$$s_D(x, y) = e^{-\beta^*|y-x|}$$

The identity relation for crisp relational databases induces equivalence classes (most frequently singleton sets) over a domain, D, which affect the results of certain operations and the removal of redundant tuples. The identity relation is replaced in this fuzzy relational database by explicitly declared similarity relations, of which an identity relation is a special case.

Next the basic concepts of fuzzy tuples and interpretations must be described. As discussed, a key aspect of most fuzzy relational databases is that domain values need not be atomic. A domain value, d_i, where i is the index of the attribute in the tuple, is defined to be a subset of its domain base set, D_i. That is, any member of the power set may be a domain value except the null set. Let $P(D_i)$ denote the power set of $D_i - \emptyset$.

A fuzzy relation R is a subset of the set cross product $P(D_1) \times P(D_2) \times \cdots \times P(D_m)$. Membership in a specific relation, r, is determined by the underlying semantics of the relation. For instance, if D_1 is the set of major cities and D_2 is the set of countries, then (Paris, Belgium) $\subset P(D_1) \times P(D_2)$—but is not a member of the relation A (capital-city, country).

A fuzzy tuple, t, is any member of both r and $P(D_1) \times P(D_2) \times \cdots \times P(D_m)$. An arbitrary tuple is of the form $t_i = (d_{i1}, d_{i2}, ..., d_{im})$ where $d_{ij} \in D_j$.

An interpretation $\alpha = [a_1, a_2, ..., a_m]$ of a tuple $t_i = (d_{i1}, d_{i2}, ..., d_{im})$ is any value assignment such that $a_j \in d_{ij}$ for all j.

In summary, the space of interpretations is the set cross product $D_1 \times D_2 \times \cdots \times D_m$. However, for any particular relation, the space is limited by the set of valid tuples. Valid tuples are determined by an underlying semantics of the relation. Note that in an ordinary relational database, a tuple is equivalent to its interpretation.

3.3   Fuzzy Database Entropy

Fuzzy entropy may be measured as a function of a domain value or as a function of a relation. Intuitively, the uncertainty of a domain value increases as its cardinality $|d_{ij}|$ increases, or when the similarity $s_j(x, y)$ decreases. So if a domain value, d_{ij}, in a relational scheme, consisting of a single element represents exact information and multiple elements are a result of fuzziness, then this uncertainty can be represented by entropy. DeLuca and Termini [23] have devised formulas for uncertainty based on fuzzy measures. Adapting their result to a fuzzy database, the entropy $H_{fz}(d_{ij})$, for a domain value $d_{ij} \subseteq D_j$ would be [2]

$$H_{fz}(d_{ij}) = - \sum_{\{x,y\} \subseteq d_{ij}} \left[s_j(x,y) log_2\left(s_j(x,y)\right) + \left(1 - s_j(x,y)\right) log_2\left(1 - s_j(x,y)\right) \right]$$

Note that $H_{fz}(d_{ij})$ is directly proportional to $|d_{ij}|$ and inversely proportional to $s_j(x,y) > 0.5$.

This definition cannot be directly extended to tuples, so a probabilistic entropy measure after Shannon is needed for an entire tuple. First recalling the interpretation of a tuple, for the ith tuple, ti, there are α_i possible interpretations, i.e., the cardinality of the cross product of the domain values, $|d_{i1} \times d_{i2} \times \cdots \times d_{im}|$. Viewing all interpretations as a priori equally likely, the entropy of tuple t_i can be given as

$$H_{pb}(t_i) = - \sum_{k=0}^{a_i} (1/ai) log_2(1/ai) = log_2(\alpha i)$$

For a nonfuzzy database, clearly $\alpha_i = 1$ and $H_{pb}(t_i) = 0$.

If the choice of a tuple in a relation r is independent of the interpretation of the tuple, the joint probabilistic entropy $H_{pb}(r, t)$ of a relation can be expressed as

$$H_{pb}(r,t) = - \sum_{i=0}^{n} \sum_{k=1}^{a_i} (n\alpha_i)^{-1} log_2\left[(n\alpha_i)^{-1} \right]$$

where there are n tuples.

Also, a query response measure can be given for a Boolean query with linguistic modifiers by using the membership value $\mu_Q(t)$ for each tuple in the relation r which is the response to a query Q. This membership value is not static but represents the best matching interpretation of the tuple t relative to the query. So the fuzzy entropy of a relation r with n tuples is

$$H_{fz}(r:Q) = - \sum_{i=1}^{n} \left[\mu_Q(t_i) log_2\left(\mu_Q(t_i)\right) + \left(1 - \mu_Q(t_i)\right) log_2\left(1 - \mu_Q(t_i)\right) \right]$$

Note that $H_{fz}(r:Q) = 0$ if and only if $(\mu_Q(t_i) = 0)$ or $(\mu_Q(t_i) = 1)$ for all i. In every other case $H_{fz}(r:Q) > 0$ and is maximized when $\mu_Q(t_i) = 0.5$ for all i. This maximization condition is achieved when a query fails to distinguish the dominant truth value of any tuple.

3.4 Rough Set Relational Model

The rough relational database model [24] as for the fuzzy database is an extension of the standard relational database model of Codd [25]. It captures all the essential features of rough set theory including indiscernibility of elements denoted by equivalence classes and lower and upper approximation regions for defining sets which are indefinable in terms of the indiscernibility. Here we relate the concepts of information theory to rough sets and compare these information theoretic measures to established rough set metrics of uncertainty. The measures are then applied to the rough relational database model. Information content of both stored relational schemas and rough relations are expressed as types of rough entropy.

3.4.1 Rough Set Theory

Rough set theory, introduced by Pawlak [26], is a technique for dealing with uncertainty and for identifying cause-effect relationships in databases. An extensive theory for rough sets and their properties has been developed and this has become a well established approach for the management of uncertainty in a variety of applications. Rough sets involve the following:

U is the universe, which cannot be empty,
R is the indiscernibility relation, or equivalence relation,
$A = (U, R)$, an ordered pair, is called an approximation space,
$[x]_R$ denotes the equivalence class of R containing x, for any element x of U,
elementary sets in A—the equivalence classes of R,
definable set in A—any finite union of elementary sets in A.

Given an approximation space defined on some universe U that has an equivalence relation R imposed upon it, U is partitioned into equivalence classes called elementary sets that may be used to define other sets in A. A rough set X, where $X \subseteq U$, can be defined in terms of the definable sets in A by the following:

lower approximation of X in A is the set $\underline{R}X = \{x \in U \mid [x]_R \subseteq X\}$
upper approximation of X in A is the set $\bar{R}X = \{x \in U \mid [x]_R \cap X \neq \emptyset\}$.

$POS_R(X) = \underline{R}X$ denotes the R-positive region of X, or those elements which certainly belong to the rough set. The R−negative region of X, $NEG_R(X) = U - \bar{R}X$, contains elements which do not belong to the rough set. The boundary or R-borderline region of X, $BN_R(X) = \bar{R}X - \underline{R}X$, contains those elements which may or may not belong to the set. X is R-definable if and only if $\underline{R}X = \bar{R}X$. Otherwise, $\underline{R}X \neq \bar{R}X$ and X is rough with respect to R. A rough set in A is the group of subsets of U with the same upper and lower approximations.

3.4.2 Rough Database Model

Every attribute domain is partitioned by some equivalence relation specified by the database designer or user. Within each domain, those values that are considered indiscernible belong to an equivalence class. This information is used by the query mechanism to retrieve information based on equivalence with the class to which the value belongs rather than equality, resulting in less critical wording of queries.

Definition A rough relation R is a subset of the set cross product $P(D_1) \times P(D_2) \times \cdots \times P(D_m)$.

A rough tuple t is any member of R, which implies that it is also a member of $P(D_1) \times P(D_2) \times \cdots \times P(D_m)$. If t_i is some arbitrary tuple, then $t_i = (d_{i1}, d_{i2}, ..., d_{im})$ where $d_{ij} \subseteq D_j$. Again a tuple in this model differs from that of ordinary databases in that the tuple components may be sets of domain values rather than single values.

Let $[d_{xy}]$ denote the equivalence class to which d_{xy} belongs. When d_{xy} is a set of values, the equivalence class is formed by taking the union of equivalence classes of members of the set; if $d_{xy} = \{c_1, c_2, ..., c_n\}$, then $[d_{xy}] = [c_1] \cup [c_2] \cup \cdots \cup [c_n]$.

Definition Tuples $t_i = (d_{i1}, d_{i2}, ..., d_{im})$ and $t_k = (d_{k1}, d_{k2}, ..., d_{km})$ are redundant if $[d_{ij}] = [d_{kj}]$ for all $j = 1, ..., m$.

Again for the rough relational database, redundant tuples are removed in the merging process since duplicates are not allowed in sets, the structure upon which the relational model is based.

3.5 Rough Set Uncertainty Metrics

Rough set theory [26] inherently models two types of uncertainty. The first type of uncertainty arises from the indiscernibility relation that is imposed on the universe, partitioning all values into a finite set of equivalence classes. If every equivalence class contains only one value, then there is no loss of information caused by the partitioning. In any coarser partitioning, however, there are fewer classes, and each class will contain a larger number of members. Our knowledge, or information, about a particular value decreases as the granularity of the partitioning becomes coarser.

Uncertainty is also modeled through the approximation regions of rough sets where elements of the lower approximation region have total participation in the rough set and those of the upper approximation region have uncertain participation in the rough set. Equivalently, the lower approximation is the certain region and the boundary area of the upper approximation region is the possible region.

Pawlak [27] discusses two numerical characterizations of imprecision of a rough set X: accuracy and roughness. Accuracy, which is simply the ratio of the number of elements in the lower approximation of X, $\underline{R}X$, to the number of elements in the upper approximation of the rough set X, $\bar{R}X$, measures the degree of completeness of knowledge about the given rough set X. It is defined as a ratio of the two set cardinalities as follows:

$$\alpha_R(X) = card(RX)/card(\bar{R}X), \text{ where } 0 \le \alpha_R(X) \le 1.$$

The second measure, roughness, represents the degree of incompleteness of knowledge about the rough set. It is calculated by subtracting the accuracy from one [10]:

$$\rho_R(X) = 1 - \alpha_R(X).$$

These measures require knowledge of the number of elements in each of the approximation regions and are good metrics for uncertainty as it arises from the boundary region, implicitly taking into account equivalence classes as they belong wholly or partially to the set. However, accuracy and roughness measures do not necessarily provide us with information on the uncertainty related to the granularity of the indiscernibility relation for those values that are totally included in the lower approximation region. For example, let the rough set X be defined as follows:

$$X = \{A11, A12, A21, A22, B11, C1\}$$

with lower and upper approximation regions defined as

$$\underline{R}X = \{A11, A12, A21, A22\} \text{ and } \bar{R}X = \{A11, A12, A21, A22, B11, B12, B13, C1, C2\}$$

These approximation regions may result from one of several partitionings. Consider, for example, the following indiscernibility relations:

$$A_1 = \{[A11, A12, A21, A22], [B11, B12, B13], [C1, C2]\},$$
$$A_2 = \{[A11, A12], [A21, A22], [B11, B12, B13], [C1, C2]\},$$
$$A_3 = \{[A11], [A12], [A21], [A22], [B11, B12, B13], [C1, C2]\}.$$

All three of the above partitionings result in the same upper and lower approximation regions for the given set X, and hence the same accuracy measure ($4/9 = 0.44$) since only those classes belonging to the lower approximation region were re-partitioned. It is obvious, however, that there is more uncertainty in A_1 than in A_2, and more uncertainty in A_2 than in A_3. Therefore, a more comprehensive measure of uncertainty is needed.

3.6 Rough Set Entropy

We derive such a measure from techniques used for measuring entropy in classical information theory. Many variations of the classical entropy have been developed, each tailored for a particular application domain or for measuring a particular type of uncertainty. Our rough entropy is defined such that we may apply it to rough databases [5]. We define the entropy of a rough set X as follows:

Definition The rough entropy $E_r(X)$ of a rough set X is calculated by

$$E_r(X) = -(\rho_R(X)) \left[\sum_{i=1}^{n} Q_i log(P_i) \right] \quad i = 1, \ldots, n \text{ equivalence classes}$$

The term $\rho_R(X)$ denotes the roughness of the set X. The second term is the summation of the probabilities for each equivalence class belonging either wholly or in part to the rough set X. There is no ordering associated with individual class members. Therefore the probability of any one value of the class being named is the reciprocal of the number of elements in the class. If c_i is the cardinality of, or number of elements in equivalence class i and all members of a given equivalence class are equal, $P_i = 1/c_i$ represents the probability of one of the values in class i. Q_i denotes the probability of equivalence class i within the universe. Q_i is computed by taking the number of elements in class i and dividing by the total number of elements in all equivalence classes combined. The entropy of the sample rough set X, $E_r(X)$, is given below for each of the possible indiscernibility relations A_1, A_2, and A_3.

Using A_1: $-(5/9)[(4/9)\log(1/4) + (3/9)\log(1/3) + (2/9)\log(1/2)] = 0.274$

Using A_2: $-(5/9)[(2/9)\log(1/2) + (2/9)\log(1/2) + (3/9)\log(1/3) + (2/9)\log(1/2)] = 0.20$

Using A_3: $-(5/9)[(1/9)\log(1) + (1/9)\log(1) + (1/9)\log(1) + (1/9)\log(1) + (3/9)\log(1/3) + (2/9)\log(1/2)] = 0.048$

From the above calculations it is clear that although each of the partitionings results in identical roughness measures, the entropy decreases as the classes become smaller through finer partitionings.

3.7 Entropy and the Rough Relational Database

The basic concepts of rough sets and their information-theoretic measures carries over to the rough relational database model. Recall that in the rough relational database all domains are partitioned into equivalence classes and relations are not restricted to first normal form. We therefore have a type of rough set for each attribute of a relation. This results in a rough relation, since any tuple having a value for an attribute that belongs to the boundary region of its domain is a tuple belonging to the boundary region of the rough relation.

There are two things to consider when measuring uncertainty in databases: uncertainty or entropy of a rough relation that exists in a database at some given time and the entropy of a relation schema for an existing relation or query result. We must consider both since the approximation regions only come about by set values for attributes in given tuples. Without the extension of a database containing actual values, we only know about indiscernibility of attributes. We cannot consider the approximation regions. We define the entropy for a rough relation schema as follows:

Definition Rough Schema Entropy

The rough schema entropy for a rough relation schema S is

$$E_s(S) = - \sum_{j=1}^{m} \left[\sum_{i=1}^{n} Q_i log(P_i) \right]$$

where there are n equivalence classes of domain j, and m attributes in the schema R $(A_1, A_2, ..., A_m)$.

This is similar to the definition of entropy for rough sets without factoring in roughness, since there are no elements in the boundary region (lower approximation = upper approximation). However, because a relation is a cross product of the domains, we must take the sum of all these entropies to obtain the entropy of the schema. The schema entropy provides a measure of the uncertainty inherent in the definition of the rough relation schema taking into account the partitioning of the domains on which the attributes of the schema are defined.

We extend the schema entropy $E_s(S)$ to define the entropy of an actual rough relation instance $E_R(R)$ of some database D by multiplying each term in the product by the roughness of the rough set of values for the domain of that given attribute.

Definition Rough Relation Entropy

The rough relation entropy of a particular extension of a schema is

$$E_R(R) = - \sum_{j=1}^{m} D\rho_j(R) \left[\sum_{i=1}^{n} DQ_i log(DP_i) \right]$$

where $D\rho_j(R)$ represents a type of database roughness for the rough set of values of the domain for attribute j of the relation, m is the number of attributes in the database relation, and n is the number of equivalence classes for a given domain for the database.

We obtain the $D\rho_j(R)$ values by letting the non-singleton domain values represent elements of the boundary region, computing the original rough set accuracy and subtracting it from one to obtain the roughness. DQ_i is the probability of a tuple in the database relation having a value from class i, and DP_i is the probability of a value for class i occurring in the database relation out of all the values which are given.

Information theoretic measures again prove to be a useful metric for quantifying information content. In rough sets and the rough relational database, this is especially useful since in ordinary rough sets Pawlak's measure of roughness does not seem to capture the information content as precisely as our rough entropy measure.

In rough relational databases, knowledge about entropy can either guide the database user toward less uncertain data or act as a measure of the uncertainty of a data set or relation. As rough relations become larger in terms of the number of tuples or attributes, the automatic calculation of some measure of entropy becomes a necessity. Our rough relation entropy measure fulfills this need.

4 Probability-Possibility Aggregation and Information Measures

Effective decision-making should be able to make use of all the available, relevant information about such aggregated uncertainty. In this section we consider quantitative measures that can be used to guide the use of aggregated uncertainty. While there are a number of possible approaches to aggregate the uncertainty information that has been gathered, here we examine uncertainty aggregation by the soft computing approach of possibilistic conditioning of probability distribution representations using the approach by Yager [28].

To formalize the problem, let V be a discrete variable taking values in a space X that has both aleatory and epistemic sources of uncertainty [29]. Let there be a probability distribution $P: X \rightarrow [0, 1]$ such that $p_i \in [0, 1]$, $\sum_{i=1}^{n} p_i = 1$ that models the aleatory uncertainty. Then the epistemic uncertainty can be modeled by a possibility distribution [30] such that $\Pi: \Xi \rightarrow [0, 1]$, where $\pi(x_i)$ gives the possibility that x_i is the value of V, where $i = 1, 2, \ldots, n$. A usual requirement here is the normality condition, $\underset{x}{Max}[\pi(x)] = 1$, that is at least one element in X must be fully possible. Abbreviating our notation so that $\pi_i = \pi(x_i)$, etc. and $\pi_i = \pi(x_i)$, etc., we have $P = \{p_1, p_2, \ldots, p_n\}$ and $\Pi = \{\pi_1, \pi_2, \ldots, \pi_n\}$.

In possibilistic conditioning, a function f dependent on both P and Π is used to find a new conditioned probability distribution such that

$$f(P, \Pi) \rightarrow \hat{P}$$

where $\hat{P} = \{\hat{p}_1, \hat{p}_2, \ldots, \hat{p}_n\}$ with $\hat{p}_i = p_i \pi_i / K$; $K = \sum_{i=1}^{n} p_i \pi_i$

A strength of this approach using conditioned probability is that it also captures Zadeh's concept of consistency between the possibility and the original probability distribution. Consistency provides an intuition of concurrence between the possibility and probability distributions being aggregated. K is identical to Zadeh's possibility-probability consistency measure [30], $C_Z(\Pi, P)$; i.e. $C_Z(\Pi, P) = K$.

4.1   *Information Measures of Conditioned Probability*

In this section we apply the Shannon and Gini measures to the original and conditioned probability distributions for the 3 possibility distribution cases shown below, and compare the measures' values [31]:

$$\text{Complete Certainty: } \Pi_{CC} = \{1, 0, \ldots, 0, 0\}$$
$$\text{Complete Uncertainty: } \Pi_{CU} = \{1, 1, \ldots, 1, 1, 1\}$$
$$\text{Intermediate Uncertainty: } \Pi_{CI} = \{1, 1, ..1, 0, 0 \ldots, 0\}$$

As both measures have increasing values with increasing uncertainty, the conditioned probability will be more informative for decision-making if it's measure value is less than for the original probability. We can see that both measures basically agree for the cases, although their specific values are in different ranges.

Case 1

For the completely certain possibility, we consider only where there is no conflict and the conditioned probability is

$$\hat{P} = \{1, 0, ..0\}$$

Then we have first for both measures with the distribution P_{cc}

$$S(\hat{P}) = G(\hat{P}) = 0 = S(P_{cc}) = G(P_{cc})$$

But for the equi-probable initial distribution P_{cu}

$$S(P_{cu}) = \ln(n) > S(\hat{P}) = 0$$
$$G(P_{cu}) = \frac{n-1}{n} > G(\hat{P}) = 0$$

So the conditioned probability distribution is more informative in the second case for the probability P_{cu}.

Case 2

Next for the case of complete possibilistic uncertainty, we have $\hat{P} = P$ for all the probability distributions and so we have

$$S(\hat{P}) = S(P) \text{ and } G(\hat{P}) = G(P)$$

We can conclude that the conditioned probability distribution $\hat{P}$ is no more informative than the original probability P since the possibility distribution Π does not contribute any information as it represents complete uncertainty.

Case 3

For the intermediate possibility case and here we consider the probability, P_{cc}, first for the Shannon measure and then the Gini index. Since for no conflict $\hat{P} = \{0, 0, ..., 0, p_t = 1, ... 0\}$ then as before for this distribution

$$S(\hat{P}) = S(P_{cc}) = 0 \text{ and } G(\hat{P}) = G(P_{cc}) = 0$$

Next for the equi-probable distribution P_{cu}, the Shannon measure is

$$S(\hat{P}) = -\left(\sum_{i=1}^{m} \left(\frac{1}{m} \right) ln\left(\frac{1}{m} \right) + \sum_{i=m+1}^{n} 0 \, ln(0) \right) = -\frac{1}{m} \sum_{i=1}^{m} (ln(1) - ln(m))$$

$$= -\frac{1}{m} * (-m \, ln(m)) = ln(m) \frac{1}{m}$$

Now since P_{cu} is an equi-probable distribution and $n > m$

$$S(P_{cu}) = ln(n) > ln(m) = S(\hat{P})$$

Next for the Gini measure

$$G(\hat{P}) = 1 - \left(\sum_{i=1}^{m} \hat{p}_i^2 + \sum_{i=m+1}^{n} \hat{p}_i^2 \right) = 1 - \left(\sum_{i=1}^{m} \left(\frac{1}{m} \right)^2 + \sum_{i=m+1}^{n} 0 \right) = 1 - m * \frac{1}{m^2}$$

Recall $G(P_{cu}) = = 1 - 1/n$ and since $1 < m < n$, $1/n < 1/m$

$$G(\hat{P}) = 1 - \frac{1}{m} < 1 - \frac{1}{n} = G(P_{cu})$$

Thus we see that by both measures the conditioned probability is more informative in this case.

5 Summary

Information measures such as Shannon and Gini have been shown to provide valuable metrics for both database quantification as well assessments of aggregation approaches. Advanced measures of information such as various parametric measures [32, 33] could also be considered for the applications described in this chapter. Additionally there are other approaches to aggregation of probability and possibility distributions being evaluated based on the transformations [34] of possibility distributions to probability distributions. Then the probability distributions can be combined and the result assessed by information measures as discussed in this chapter [35].

Acknowledgments We would like to thank the Naval Research Laboratory's Base Program, Program Element No. 0602435N for sponsoring this research.

References

1. Shannon, C.: A mathematical theory of communication. Bell Syst. Tech. J. **27**(379–423), 623–656 (1948)
2. Buckles, B., Petry, F.: Information-theoretical characterization of fuzzy relational databases. IEEE Trans. Syst. Man Cybern. **13**, 74–77 (1983)
3. Fung, K.T., Lam, C.M.: The database entropy concept and its application to the data allocation problem. INFOR **18**, 354–363 (1980)
4. Quinlan, J.: Induction of decision trees. Mach. Learn. **1**, 81–106 (1986)
5. Beaubouef, T., Petry, F., Arora, G.: Information-theoretic measures of uncertainty for rough sets and rough relational databases. Inf. Sci. **109**, 185–195 (1998)
6. Yovits, M., Foulk, C.: Experiments and analysis of information use and value in a decision-making context. J. Am. Inf. Sci. **36**(2), 63–81 (1985)
7. Klir, G.: Uncertainty and Information. John Wiley, Hoboken NJ (2006)
8. Reza, F.: An Introduction to Information Theory. McGraw Hill, New York (1961)
9. Gini, C.: Variabilita e mutabilita (Variability and mutability), Tipografia di Paolo Cuppini, Bologna, Italy, p. 156 (1912)
10. Aristondo, O., Garcia-Lparesta, J., de la Vega, C., Pereira, R.: The Gini index, the dual decomposition of aggregation functions and the consistent measurement of inequality. Int. J. Intell. Syst. **27**, 132–152 (2012)
11. Breiman, L., Friedman, J., Olshen, R., Stone, C.: Classification and Regression Trees. Wadsworth & Brooks/Cole, Monterey, CA (1984)
12. Sent, D., van de Gaag, L.: On the behavior of information measures for test selection. In: Carbonell, J., Siebnarm, J. (eds.) Lecture Notes in AI 4594, pp. 325–343. Springer, Berlin (2007)
13. Eliazar, I., Sokolov, I.: Maximization of statistical heterogeneity: from Shannon's entropy to Gini's index. Phys. A **389**, 3023–3038 (2010)
14. Yager, R.: Measures of entropy and fuzziness related to aggregation operators. Inf. Sci. **82**, 147–166 (1995)
15. Renyi, A.: On measures of information and entropy. In: Proceedings of the 4th Berkeley Symposium on Mathematics, Statistics and Probability 1960, pp. 547–561 (1961)
16. Renyi, A.: Probability Theory. North-Holland, Amsterdam, ND (1970)
17. Hartley, R.: Transmission of information. Bell Syst. Tech. J. **7**(3), 535–563 (1928)
18. Ola, A., Ozsoyoglu, G.: Incomplete relational database models based on intervals. IEEE Trans. Knowl. Data Eng. **5**, 293–308 (1993)
19. Roth, M., Korth, H., Batory, D.: SQL/NF: A query language for non-1NF databases. Inf. Syst. **12**, 99–114 (1987)
20. Petry, F.: Fuzzy Databases: Principles and Applications. Kluwer Press, Norwell MA (1996)
21. Buckles, B., Petry, F.: A fuzzy representation for a relational data base. Int. J. Fuzzy Sets Syst. **7**, 213–226 (1982)
22. Zadeh, L.: Similarity relations and fuzzy orderings. Inf. Sci. **3**(2), 177–200 (1971)
23. de Luca, A., Termini, S.: A definition of a nonprobabilistic entropy in the setting of fuzzy set theory. Inf. Control **20**, 301–312 (1972)
24. Beaubouef, T., Petry, F., Buckles, B.: Extension of the relational database and its algebra with rough set techniques. Comput. Intell. **11**, 233–245 (1995)
25. Elmasri, R., Navathe, S.: Fundamentals of Database Systems, 4th edn. Addison Wesley, New York (2004)
26. Pawlak, Z.: Rough sets. Int. J. Comput. Inf. Sci. **11**, 341–356 (1982)

27. Pawlak, Z.: Rough Sets: Theoretical Aspects of Reasoning about Data. Kluwer, Norwell, MA (1991)
28. Yager, R.: Conditional approach to possibility-probability fusion. IEEE Trans. Fuzzy Syst. **20** (1), 46–56 (2012)
29. Parsons, S.: Qualitative Methods for Reasoning Under Uncertainty. MIT Press, Cambridge, MA (2001)
30. Zadeh, L.: Fuzzy sets as a basis for a theory of possibility. Fuzzy Sets Syst. **1**, 3–35 (1978)
31. Elmore, P., Petry, F., Yager, R.: Comparative measures of aggregated uncertainty representations. J. Ambient Intell. Humanized Comput. **5**, 809–819 (2014)
32. Arora, G., Petry, F., Beaubouef, T.: Uncertainty measures of type B under similarity relations. Int. J. Pure Appl. Math. **2**, 219–233 (2002)
33. Arora, G., Petry, F., Beaubouef, T.: A note on parametric measures of information for fuzzy sets, Int. J. Comb. Inf. Syst. Sci. **26**, 167–174 (2003)
34. Dubois, D., Prade, H.: On several representations of an uncertain body of evidence. In: Gupta, M., Sanchez, E. (eds.) Fuzzy Information and Decision Processes, pp. 167–182. North Holland, Amsterdam (1982)
35. Petry, F., Elmore, P., Yager, R.: Combining uncertain information of differing modalities. Inf. Sci. **322**, 237–256 (2015)

Part II
Applications in Modeling, Decision Making, Control, and Other Areas

On Practical Applicability of the Generalized Averaging Operator in Fuzzy Decision Making

Uzay Kaymak

Abstract Many different types of aggregation operators have been suggested as decision functions for multicriteria fuzzy decision making. This paper investigates the practical applicability of generalized averaging operator as decision functions in modeling human decision behavior. Previously published numerical data is used in the analysis and the results are compared with those obtained from compensatory operators. The numerical data suggests that the generalized averaging operator may be used for modeling human decision behavior.

1 Introduction

Decision making in a fuzzy environment has been introduced in [2] as the intersection of fuzzy sets representing the goals and the constraints of the decision problem. The minimum operator was used to determine the intersection of fuzzy sets. Subsequent research has shown that other aggregation operators from the fuzzy set theory can also be used to formulate the aggregation of information concerning the goals and the constraints. In this respect, a lot of attention has been paid to fuzzy aggregation operators and the aggregation of information modeled by fuzzy sets. Consequently, the field of aggregation operators is recognized as an important cornerstone of fuzzy systems research (see e.g. [1, 5, 6, 9, 15, 22]).

One of the important contributors to the field has been Ron Yager, who has introduced weighted aggregation of fuzzy information [25], informational considerations of aggregation [26] and new aggregation operators, such as Yager's t-norm [27] and the ordered weighted averaging operators [28]. Nowadays, it is widely accepted that fuzzy decision making can use any appropriate aggregation of fuzzy goals and constraints, based one of the many fuzzy set aggregation operators. The selection of a suitable decision function for a problem then reflects the aims of the decision maker.

U. Kaymak (✉)
School of Industrial Engineering, Eindhoven University of Technology,
P.O. Box 513, 5600 MB Eindhoven, The Netherlands
e-mail: u.kaymak@ieee.org

© Springer International Publishing Switzerland 2017   101
J. Kacprzyk et al. (eds.), *Granular, Soft and Fuzzy Approaches*
for Intelligent Systems, Studies in Fuzziness and Soft Computing 344,
DOI 10.1007/978-3-319-40314-4_6

The decision maker may choose a decision function that suits his purposes best. Furthermore, certain decision functions may be more suitable for certain types of decision problems. Hence, the selection of a suitable decision function involves some uncertainty. Consequently, many possible decision functions have been suggested in literature. The most common ones are the t-norms for modelling the conjunctive (and-type) aggregation of the criteria and the t-conorms for modelling the disjunctive (or-type) aggregation. Even though there has been an abundance of proposed decision functions, the practical applicability of these functions in human decision making has been empirically tested and shown only for a small number of them. Studies carried out by Thole [23], Kovalerchuk [17] and Zimmermann [33] indicate that the most commonly used t-norms (minimum and product operator) and their associated t-conorms are not suitable for modelling human decision behaviour. It seems that human beings do not aggregate criteria by t-norms or t-conorms alone, but by a compensatory combination of the criteria. A general form of compensatory operators has been defined by Mizumoto [21]. Some special compensatory operators have been suggested and studied empirically by Zimmermann [33]. The numerical results show that the investigated compensatory operators approximate the human decision behaviour sufficiently well.

Another class of aggregation operators that allows compensation between criteria is the quasi-linear averaging operators that have been studied by van Nauta Lemke et al. [24] and by Dyckhoff and Pedrycz [8] in the fuzzy set setting. Although common in many areas of decision making, it is an open question whether the averaging operators can be used to model the way humans aggregate information. In this sense, thorough empirical studies are still necessary in order to assess to what extent generalized averaging operators model the human decision behaviour. In this paper we want to give impulse for more empirical investigation of the use of averaging operators in fuzzy decision making, by comparing empirical results obtained with the averaging operators with those obtained from the compensatory operators as suggested by Zimmermann and Zysno [33]. To facilitate the comparison, we use previously published empirical data of Thole et al. [23], Kovalerchuk et al. [17] and Zimmermann et al. [33].

The paper is organized as follows. Section 2 gives a summary of different types of decision functions and gives a brief introduction to the decision functions used in our study. An interpretation of various parameters used in these decision functions is given. Section 3 summarizes the empirical studies conducted by Thole et al. [23], Kovalerchuk et al. [17] and Zimmermann et al. [33], whose numerical results have been used in this paper. It also lists the conditions upon which the suitability of the generalized averaging operator for decision making has been tested. Section 4 gives the numerical results obtained through the use of the averaging operators and compares them to the results obtained from the compensatory operators described in [33]. The paper ends with the conclusions in Sect. 5.

2 Decision Functions

As discussed in Sect. 1, decision making in a fuzzy environment has been defined by Bellman and Zadeh [2] as the intersection of fuzzy sets representing the goals and/or the constraints of the decision problem, but it is widely accepted, nowadays, that any suitable aggregation of fuzzy sets may be used in fuzzy decision making. Consequently, many aggregation operators have been proposed in the literature [2, 6–8, 21, 24, 25, 28, 29, 31, 33]. Because the decision is made as a result of this aggregation, the functions which combine a number of fuzzy sets by using these operators are known as decision functions.

Since different decision makers will have different aims in a decision problem, it is expected that the fuzzy sets corresponding to the goals and the constraints may be aggregated in different ways. The decision maker may choose a decision function that best reflects the goals of the decision. Example 1 illustrates this.

Example 1 A young man is going to buy a bunch of flowers for his girl friend. He may wish the flowers to be brightly coloured *and* smell nice. Another possibility is that he wishes them to be brightly coloured *or* smell nice. Clearly, he has to use different decision functions in each case (e.g. minimum operator as opposed to the maximum operator).

It may also be the case that certain decision functions may inherently be more suitable for certain types of decision problems. In this case, it is the boundary conditions on the decision, rather than the personal preferences of the decision maker that determines the choice of the decision function.

Example 2 A person wants to buy a car that is fast, economical (low fuel usage) and inexpensive. It is probably not possible to satisfy all these criteria at the same time. A fast car will not be economical and an inexpensive car probably will not be fast. Hence, one is forced to make a trade-off between the criteria. Therefore, the decision function must be chosen so as to allow for trade-off between criteria.

2.1 Common Aggregation Types

Clearly, the selection of a decision function is of central importance in the fuzzy decision making model. If the decision maker chooses to satisfy all the criteria simultaneously, conjunction operators are used as decision functions. When the decision maker allows full compensation between criteria (i.e. he is satisfied when at least one criterion is satisfied), disjunction operators are used. T-norms and t-conorms are used as conjunction and disjunction operators respectively [3]. For an axiomatic definition of t-norms and t-conorms, the reader may refer to [6, 16, 20]. We want to mention here one property of t-norms and t-conorms that is of relevance in empirical studies:

1. Minimum operator is the largest t-norm, i.e. $T(a, b) \leq \min(a, b)$, $a, b \in [0, 1]$;
2. Maximum operator is the smallest t-conorm, i.e. $S(a, b) \geq \max(a, b)$, $a, b \in [0, 1]$.

There has been indications in the literature [23, 30] that this property of t-norms and t-conorms make them unsuitable for the modelling of aggregation by human decision makers. It appears that human beings tend to partially compensate between criteria instead of trying to satisfy them simultaneously or make complete compensations. According to one interpretation, human beings use a mixture of conjunction and disjunction in their decisions. To model this, compensatory operators have been proposed, the general form of which has been defined by [21]:

$$\hat{C}(a, b) = M (F(a, b), G(a, b)) \tag{1}$$

where $M(a, b)$ is an averaging operator, and $F(a, b)$ and $G(a, b)$ are t-norms, t-conorms or averaging operators. Special forms of (1) have been suggested and investigated by Zimmermann and Zysno [33]. These have the form:

$$\mu_{A \ominus B} = \left(\mu_{A \cap B} \right)^{1-\gamma} \cdot \left(\mu_{A \cup B} \right)^{\gamma} \tag{2}$$

and

$$\mu_{A \ominus B} = (1 - \gamma)\mu_{A \cap B} + \gamma \mu_{A \cup B} \tag{3}$$

where γ is interpreted as the 'grade of compensation'. Different decision functions may now be obtained by the suitable selection of γ and the intersection and the union operator for the fuzzy sets. Usually the minimum and the maximum operators are used for the intersection and the union respectively. However, [33] suggests the use of the algebraic product and the algebraic sum as these operators allow interaction between fuzzy sets.

2.2 Generalized Averaging Operator

Another class of operators that allows compensation between criteria is the averaging operators. Averaging operators satisfy the following conditions:

1. $a \wedge b \leq M(a, b) \leq a \vee b$,
2. $M(a, b) = M(b, a)$,
3. $a \leq c \iff M(a, b) \leq M(c, b)$.

Averaging operators are natural operators for decision problems where there is a trade-off between criteria, because the solution is always intermediate between the best and the worst satisfied criteria.

Averaging operators are quite common in various fields of decision making. Minimum operator has been used in economics, for example, while arithmetic mean is used, amongst others, in education. Quadratic mean is used in control theory and lies at the basis of the least squares method. Harmonic mean is sometimes used while the product operators are usually employed in probability analysis.

When all the criteria are equally important, these commonly used averaging operators, together with many others, can be brought under a parametric formula, which is

$$D_i(s) = \left\{ \frac{1}{m} \sum_{j=1}^{m} \mu_{ij}^s \right\}^{1/s} \qquad s \in \mathbb{R} \backslash \{0\}, \qquad \mu_{ij} \in [0, 1] \qquad (4)$$

with

$$D_i(0) \triangleq \prod_{j=1}^{m} \mu_{ij}^{1/m}. \qquad (5)$$

Herein s is a parameter that may take any real value, m is the total number of criteria considered, while μ_{ij} is the degree with which alternative a_i satisfies the criterion c_j. $D_i(s)$ is the overall evaluation for alternative a_i. By changing the value of the parameter s, one obtains different decision functions, such as the minimum (for $s \to -\infty$), harmonic mean (for $s = -1$), geometric mean (for $s = 0$), arithmetic mean (for $s = 1$), quadratic mean (for $s = 2$) or the maximum (for $s \to \infty$).

Equation (4) satisfies a number of general properties.

1. $D_i(s)$ is a continuous function of parameter s.
2. $D_i(s)$ is monotonic and non-decreasing as a function of s.
3. Provided that $\mu_{ij} \in [0, 1]$, $D_i(s) \in [0, 1]$.
4. $D_i(s)$ is an increasing function of μ_{ij}.

The proofs for property 1 and property 2 can be found in [8, 11]. Property 3 follows from the boundary conditions (minimum and maximum) and property 1. Property 4 is a necessary condition for (4) to be an averaging operator and has been studied in [12] in the fuzzy sets context.

The parameter s allows the generalized mean to be customized for a given setting (preference of the decision maker). In the literature, it has been shown that s can be interpreted as a characteristic index of optimism of the decision maker [13, 24]. By changing the value of the parameter s, the decision function can be adapted to the context, so that it fits the decision problem as good as possible. This adaptation also corresponds to determining the characteristic degree of optimism of the decision maker.

2.3 Hurwicz Criterion

The optimism-pessimism criterion of Hurwicz is sometimes used in the decision making literature [14, 19]. This is a decision function with an optimism index σ. Hurwicz criterion is given in (6).

$$D_H = (1 - \sigma) \bigwedge_{j=1}^{m} \mu_{ij} + \sigma \bigvee_{j=1}^{m} \mu_{ij} \qquad \sigma \in [0, 1]. \tag{6}$$

For $\sigma = 0$ one obtains the minimum operator and for $\sigma = 1$ one obtains the maximum operator. The disadvantage of this criterion and σ as an index of optimism is that it depends only on the minimum and the maximum of the membership values. The criterion is insensitive to membership values between the two extremes. The decision function (4) with s as the index of optimism does not suffer from this disadvantage. Equation (4) makes it possible that all the membership values contribute towards the value of the decision function in amounts that are determined, amongst others by the value of the optimism index.

Note that the Hurwicz criterion is a linear combination of the minimum and the maximum of the membership values. It is a special case of (3) with

$$\mu_\cap = \bigcap_{j=1}^{m} \mu_{ij} = \bigwedge_{j=1}^{m} \mu_{ij}, \tag{7}$$

and

$$\mu_\cup = \bigcup_{j=1}^{m} \mu_{ij} = \bigvee_{j=1}^{m} \mu_{ij}. \tag{8}$$

2.4   Zimmermann Operator

In [33], the authors propose to use (2) with

$$\mu_\cap = \bigcap_{j=1}^{m} \mu_{ij} = \prod_{j=1}^{m} \mu_{ij}. \tag{9}$$

and

$$\mu_\cup = \bigcup_{j=1}^{m} \mu_{ij} = 1 - \prod_{j=1}^{m} (1 - \mu_{ij}). \tag{10}$$

The product operator and the algebraic sum are preferred as the t-norm and the t-conorm, respectively, because, these operators allow for interaction amongst the criteria. In this case one obtains the following decision function:

$$D_Z = \left(\prod_{j=1}^{m} \mu_{ij} \right)^{1-\gamma} \left(1 - \prod_{j=1}^{m} (1 - \mu_{ij}) \right)^{\gamma}. \tag{11}$$

Therefore, Zimmermann prefers a logarithmical linear combination. Note that, these operators take a weighted average of a t-norm and a t-conorm. Hurwicz takes the arithmetic mean while Zimmermann takes the geometric mean. Zimmermann interprets the parameter γ as the grade of compensation between the criteria. When $\gamma = 0$, there is no compensation between criteria and one obtains a t-norm. When $\gamma = 1$, there is full compensation between the criteria and one obtains a t-conorm. The grade of compensation is similar to the optimism index but they are not equivalent. The optimism index is an indication of the decision maker's bias towards highly or badly satisfied criteria while the grade of compensation is an indication of how much compensation there should be between various (linguistic) combinations of criteria. In other words, the grade of compensation indicates the importance that the decision maker attaches to a conjunctive or disjunctive combination of criteria.

Using (6) or (11) is equivalent to using (4) with a t-norm and a t-conorm as its inputs, i.e.

$$\tilde{D} = \left\{ (1 - \gamma)\mu^s_{t-norm} + \gamma\mu^s_{t-conorm} \right\}^{1/s}. \tag{12}$$

By choosing s equal to 1 in (12), one obtains the Hurwicz criterion and by choosing s equal to 0, one obtains the Zimmermann aggregation.

3 Empirical Studies

In this section we describe the experiments conducted by Thole et al. [23], Kovaler-chuk et al. [17] and Zimmermann et al. [33] (designated respectively by *Thole79*, *Koval92* and *Zimmer80* in the rest of the paper) and reproduce their numerical results. We formulate the conditions which must be satisfied for accepting that the considered decision function can model human decision behaviour. The results are later used for testing the practical suitability of the generalized averaging operator in decision making and they are compared with the results obtained by using (6) and (11) as decision functions. The experimental studies consider conjunctive aggregation and they are designed for determining which aggregation operators are suitable for modelling the conjunctive aggregation as perceived by humans.

3.1 Experimental Design

The experiments from [17, 23, 33] have been used by various authors to demonstrate the power of various decision functions for modeling the aggregation behavior in human decision making [10, 18]. Hurwicz' optimism-pessimism operator and the generalized averaging operator have not been studied in this context, yet. We provide this analysis in the following based on the three sets of experiments.

In *Thole79* the subjects were presented with 20 different objects and asked to what degree they considered the objects to be

1. metallic,
2. container,
3. metallic container.

The objects were selected such that, they represented the membership grades from 0 to 1 rather evenly. Corrections were made for possible experimental errors (distortion, end-effect etc.). For a detailed explanation of the experimental setup the reader should refer to [23].

In *Koval92* the subjects were asked to what extent they considered various objects to be

1. heavy,
2. balls,
3. heavy balls.

For a detailed explanation of the experiment, the reader may refer to [17].

Because of the fact that an object is a metallic container when it is *both* metallic *and* container, or it is a heavy ball when it is *both* heavy *and* a ball, these experiments are expected to model an and-type combination of arguments with no compensation between them.

In *Zimmer80* the subjects had to consider a number of tiles by their

1. dovetailing,
2. solidity (indicated by a light grey colour) and
3. quality.

The quality of the tiles is determined *both* by dovetailing *and* by their solidity. However, a good dovetailing can compensate poor solidity up to a point and vice versa. Hence, compensation between criteria is possible. The reader may refer to [33] for a detailed description of the experiment. Table 1 shows the membership values measured in the three experiments.

3.2 Evaluating the Goodness of Fit

Zimmermann proposes eight criteria that can be used for selecting a decision function [32]. These include *axiomatic strength, empirical fit, adaptability, numerical efficiency, compensation, range of compensation, aggregating behavior* and *required scale level of membership functions*. The criteria describe different aspects that play a role in the selection of decision functions, but they are not totally independent of one another. In this paper, we focus on the empirical fit and are interested in which operators are able to reproduce the empirical results obtained from the aggregation of humans. For this purpose, the measured membership values for items (1) and (2)

Table 1 Empirically determined grades of membership in *Thole79*, *Koval92* and *Zimmer80*

Stimulus x	Thole79			Koval92			Zimmer80		
	$\mu_M(x)$	$\mu_C(x)$	$\mu_{M\cap C}(x)$	$\mu_H(x)$	$\mu_B(x)$	$\mu_{H\cap B}(x)$	$\mu_D(x)$	$\mu_S(x)$	$\mu_{D\cap S}$
1	0.000	0.985	0.007	0.225	0.864	0.275	0.241	0.426	0.215
2	0.908	0.419	0.517	0.256	0.225	0.123	0.662	0.352	0.427
3	0.215	0.149	0.170	0.015	0.788	0.130	0.352	0.109	0.221
4	0.552	0.804	0.674	0.081	0.010	0.015	0.052	0.630	0.212
5	0.023	0.454	0.007	1.000	0.985	0.938	0.496	0.484	0.486
6	0.501	0.437	0.493	0.069	0.656	0.115	0.000	0.000	0.000
7	0.629	0.400	0.537	0.363	0.046	0.139	0.403	0.270	0.274
8	0.847	1.000	1.000	0.935	0.301	0.408	0.130	0.156	0.119
9	0.424	0.623	0.460	0.096	0.441	0.114	0.284	0.790	0.407
10	0.318	0.212	0.142	0.203	0.746	0.229	0.193	0.725	0.261
11	0.481	0.310	0.401	0.215	0.684	0.235	1.000	1.000	1.000
12	1.000	0.000	0.000	0.350	0.383	0.240	0.912	0.330	0.632
13	0.663	0.335	0.437	0.263	0.536	0.345	0.020	0.949	0.247
14	0.283	0.448	0.239	0.000	1.000	0.125	0.826	0.202	0.500
15	0.130	0.512	0.101	0.445	0.026	0.083	0.551	0.744	0.555
16	0.325	0.239	0.301	0.871	0.056	0.169	0.691	0.572	0.585
17	0.969	0.256	0.330	0.176	0.961	0.300	0.975	0.041	0.355
18	0.480	0.012	0.023	0.251	0.383	0.205	0.873	0.534	0.661
19	0.546	0.961	0.714	0.673	0.054	0.168	0.587	0.674	0.570
20	0.127	0.980	0.185	0.203	0.710	0.290	0.450	0.440	0.418
21							0.750	0.909	0.789
22							0.091	0.856	0.303
23							0.164	0.974	0.515
24							0.788	0.073	0.324

are combined together, for each experiment, using a number of operators. The values calculated by these operators are then compared with the observed membership values for item (3).

An aggregation operator is not to be rejected as a suitable aggregation operator that models human decision behavior, if it satisfies the following two conditions [23]:

1. the mean of the difference between the observed and the calculated values ($\mu_{M \cap C}$, $\mu_{H \cap B}$ and $\mu_{D \cap S}$) is not significantly different from zero (Student's t-test with $\alpha = 0.025$, two-tailed),
2. the correlation between the observed and the calculated values is similar to 1.

In accordance with [23], we use a crisp threshold of 0.95, and require that the correlation between the observed values and the calculated values is greater than or equal to 0.95. The selection of 0.95 as the threshold value is arbitrary, but it is chosen here since it is one of the most commonly used thresholds in statistics. Another possibility could be defining a membership function which shows the level of acceptance for an operator as a function of the correlation. In that case there is a gradual transition from the non-acceptable correlation values to the acceptable correlation values.

We should make here a remark concerning condition 1. An important assumption for condition 1 is that all the means that are considered are distributed normally. The assumption has indeed been tested in a pre-test [23]. Further, if a hypothesis cannot be rejected by Student's t-test, it does not mean automatically that it *can* be accepted. However, if the t-test does not reject an operator and the correlation between the observed and the calculated values is larger than 0.95, we will consider the operator to be acceptable.

4   Numerical Results

It has been shown in [23] that, the minimum and the product operators are not suitable for modelling human decision behaviour (in connection with the *and* operator). We consider the decision function (4) in this analysis, and compare the results with those obtained by the use of Hurwicz criterion (6) and Zimmermann connective (11) to see whether they can model human decision behaviour. We investigate whether the conditions specified in the previous section are satisfied for some value of the parameter corresponding to each decision function. In order to see whether the decision functions satisfy the conditions for some value of their corresponding parameter, we plotted the graphs of the parameter t of the t-test and the correlation coefficient as a function of the parameters of the decision functions. This was done for all sets of data.

Figure 1 shows the results for the optimism index s. Note that below $t = 2$, the null-hypothesis cannot be rejected and hence condition 1 is satisfied. Condition 2 is satisfied when the value of the correlation coefficient exceeds 0.95. It is seen from the figure that it is possible to find values for the index of optimism for which both

Fig. 1 Parameter *t* and the
correlation coefficient
plotted as a function of the
index of optimism *s*

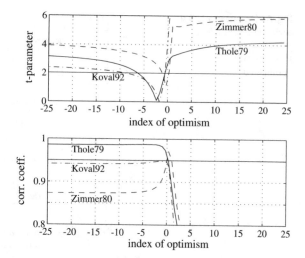

conditions are satisfied simultaneously. Hence, provided that one can find a satisfactory method for the correct determination of the decision maker's optimism, (4)
can be used for modelling human decision behaviour. Similar results may also be
obtained by the Hurwicz criterion (6) and the Zimmermann connective (11).

Figure 2 shows the results for (11), while Fig. 3 depicts the results for (6). Thus,
the compensatory operators are adequate tools for modelling human decision behaviour.

Having concluded that the averaging operators and the compensatory operators
(as expected) may be used for modelling human decision behaviour, one may wonder

Fig. 2 Parameter *t* and the
correlation coefficient
plotted as a function of the
grade of compensation *γ*

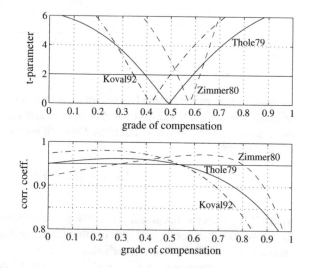

Fig. 3 Parameter t and the correlation coefficient plotted as a function of Hurwicz index of optimism σ

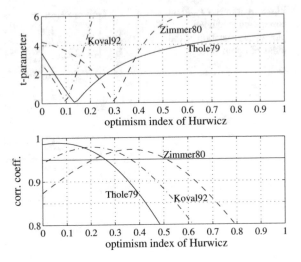

Fig. 4 Values calculated by $D_i(0.358)$ versus the observed values for *Zimmer80*. The *solid line* is the line of perfect fit

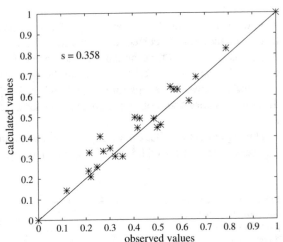

for which value of the parameters one obtains the 'best' fit such that the conditions of Sect. 3 are satisfied. The optimal choice for the parameters may be determined by the minimization of a suitable criterion. One commonly used criterion is the sum of squared errors

$$LS = \sum_{i=1}^{N} (Y_i^p - Y_i)^2 \tag{13}$$

where Y_i^p denotes the calculated values using an aggregation operator, and Y_i denotes the observed values for the and-type combination. For *Zimmer80* the optimal s-value becomes $s_{opt} = 0.358$. Even though this value minimizes the squared errors, the mean of the difference between the observed and the calculated values is signifi-

Table 2 Optimal parameter values obtained with data from *Thole79*, *Koval92* and *Zimmer80*

Stim.	Thole79				Koval92				Zimmer80			
	$\mu_{obs.}$	$D_i^*(s)$	D_Z^*	D_H^*	$\mu_{obs.}$	$D_i^*(s)$	D_Z^*	D_H^*	$\mu_{obs.}$	$D_i^*(s)$	D_Z^*	D_H^*
1	0.007	0.000	0.000	0.132	0.275	0.320	0.374	0.275	0.215	0.322	0.274	0.300
2	0.517	0.511	0.588	0.485	0.123	0.239	0.136	0.227	0.427	0.485	0.467	0.451
3	0.170	0.171	0.098	0.158	0.130	0.023	0.072	0.076	0.221	0.199	0.152	0.187
4	0.674	0.633	0.626	0.586	0.015	0.015	0.006	0.016	0.212	0.193	0.182	0.236
5	0.007	0.029	0.064	0.081	0.938	0.992	0.991	0.986	0.486	0.490	0.459	0.488
6	0.493	0.465	0.387	0.446	0.115	0.102	0.145	0.115	0.000	0.000	0.000	0.000
7	0.537	0.467	0.431	0.431	0.139	0.068	0.065	0.071	0.274	0.330	0.280	0.312
8	1.000	0.911	0.917	0.868	0.408	0.418	0.475	0.351	0.119	0.142	0.089	0.138
9	0.460	0.487	0.444	0.451	0.114	0.138	0.122	0.123	0.407	0.479	0.482	0.445
10	0.142	0.245	0.169	0.226	0.229	0.287	0.309	0.246	0.261	0.381	0.375	0.363
11	0.401	0.361	0.300	0.333	0.235	0.299	0.296	0.252	1.000	1.000	1.000	1.000
12	0.000	0.000	0.000	0.134	0.240	0.365	0.255	0.353	0.632	0.555	0.580	0.516
13	0.437	0.405	0.404	0.379	0.345	0.339	0.273	0.285	0.247	0.162	0.180	0.316
14	0.239	0.331	0.267	0.305	0.125	0.000	0.000	0.079	0.500	0.417	0.429	0.401
15	0.101	0.163	0.187	0.181	0.083	0.039	0.056	0.059	0.555	0.641	0.638	0.613
16	0.301	0.269	0.187	0.251	0.169	0.084	0.169	0.120	0.585	0.629	0.621	0.610
17	0.330	0.320	0.478	0.352	0.300	0.256	0.357	0.238	0.355	0.223	0.251	0.339
18	0.023	0.015	0.048	0.075	0.205	0.299	0.201	0.261	0.661	0.685	0.698	0.642
19	0.714	0.650	0.708	0.602	0.168	0.081	0.129	0.103	0.570	0.629	0.620	0.615
20	0.185	0.160	0.334	0.241	0.290	0.286	0.296	0.243	0.418	0.445	0.407	0.443

(continued)

Table 2 (continued)

Stim.	Thole79				Koval92				Zimmer80			
	$\mu_{obs.}$	$D_i^*(s)$	D_Z^*	D_H^*	$\mu_{obs.}$	$D_i^*(s)$	D_Z^*	D_H^*	$\mu_{obs.}$	$D_i^*(s)$	D_Z^*	D_H^*
21									0.789	0.826	0.839	0.801
22									0.303	0.295	0.312	0.335
23									0.515	0.414	0.453	0.422
24									0.324	0.255	0.262	0.301
Opt. prm.		−3.02	0.478	0.134		−1.70	0.429	0.079		0.086	0.575	0.319
LS		0.048	0.123	0.127		0.102	0.064	0.060		0.107	0.069	0.077
LQ		0.000	0.000	0.000		0.000	0.000	0.000		0.000	0.000	0.000

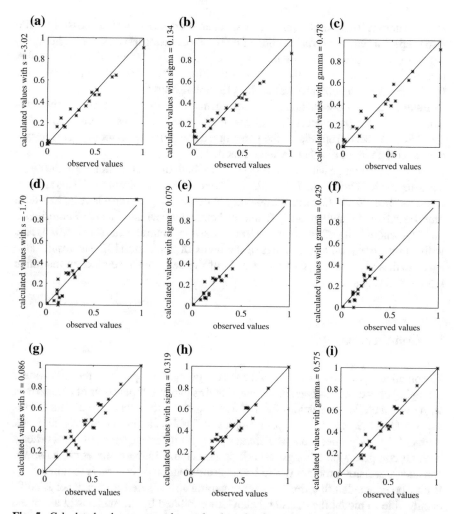

Fig. 5 Calculated values versus observed values for the optimal values of the parameters. **a** Thole79, **b** Thole79, **c** Thole79, **d** Koval92, **e** Koval92, **f** Koval92, **g** Zimmer80, **h** Zimmer80 and **i** Zimmer80

cantly different from zero, as Fig. 4 shows. This has also been tested with a *t*-test. Hence, the sum of squared errors is not a good criterion for estimating *s*. Given the conditions of Sect. 3.2, we have then chosen

$$LQ = \left| \sum_{i=1}^{N} \left(Y_i^p - Y_i \right) \left| Y_i^p - Y_i \right| \right| \tag{14}$$

as the criterion because, it minimizes the squared errors as much as possible while also keeping a symmetric distribution of points about the observed values. Therefore, condition 1 and condition 2 are both represented in this criterion. One now obtains results which satisfy both of the specified conditions. Table 2 summarizes the results. For comparison, the obtained values for (13) are also given.

Figure 5 depicts the calculated values against the observed values for the optimal values of the parameters. The diagonal line $y = x$ indicates where ideal results should lie. As seen from Fig. 5, the generalized averaging operator (4) gives results comparable to those of other decision functions.

Note that, the range within which (4) satisfies both of the conditions is in the negative s-region for *Thole79* and *Koval92*, and around $s = 0$ for *Zimmer80*. This is interesting as *Thole79* and *Koval92* are assumed to describe a problem where both criteria must be satisfied simultaneously (cautious behaviour), while *Zimmer80* describes a problem where trade-off between criteria is possible (neutral behaviour). This agrees with an interpretation of the values of the index of optimism [13]. The results also agree with a study which shows that 98 % of the decisions take place in the range $s \in [-25, 25]$ [4].

5   Conclusions

The practical applicability of a special class of aggregation operators, the quasi-linear averaging operators, has been investigated in this paper. For purposes of comparison with several earlier suggestions for decision operators, the experimental data of previous studies [17, 23, 33] have been employed. Thole et al. in [23] and Kovalerchuk et al. in [17] study a decision problem in which a conjunctive aggregation of criteria with no compensation amongst criteria is expected. Zimmermann et al., however, consider a decision problem where compensation amongst criteria is required [33]. It may be concluded that the averaging operator investigated in this paper is sufficiently able to model the decision behaviour exhibited by the subjects in all cases. The parameter s of the averaging operator may be interpreted as the characteristic optimism index of the decision maker and needs to be determined separately. We have observed that the optimal parameter value of s is in a range that corresponds to what can be expected from the type of aggregation.

Another class of operators that have been shown to model human decision behaviour sufficiently well are the compensatory operators such as those suggested by Zimmermann and Hurwicz [33]. These are actually weighted averages of a conjunction operator and a disjunction operator. The criteria only have influence on the value of the decision function through the conjunction and the disjunction operators. The membership values do not have direct influence on the value of the decision function as is the case with the averaging operators. This points to an advantage of the averaging operators as the influence of a certain criterion on the decision function may be studied relatively easily, without first taking its interaction into account, with other criteria through the conjunction and disjunction operators.

This paper gives an indication that the averaging operator (4) may be used for modelling human decision behaviour. However, more empirical studies are necessary for fully investigating the scope of its practical applicability. We hope that the results presented in this paper will stimulate researchers to further investigate the practical applicability of the averaging operators in decision making.

References

1. Beliakov, G., Pradera, A., Calvo, T.: Aggregation Functions: A Guide for Practitioners. Springer, Berlin (2007)
2. Bellman, R.E., Zadeh, L.A.: Decision-making in a fuzzy environment. Manag. Sci. 17(4), 141–164 (1970)
3. da Costa Sousa, J.M., Kaymak, U.: Fuzzy Decision Making in Modeling and Control. World Scientific Series in Robotics and Intelligent Systems, vol. 27. World Scientific, New Jersey (2002)
4. de Herder, W.A.W.: Multicriteria beslissen met vage criteria. Master's thesis report, Delft University of Technology, Control Lab. Fac. of El. Eng. Delft (1986). (A86.005–(364))
5. Dombi, J.: Basic concepts for the theory of evaluation: the aggregative operator. Eur. J. Oper. Res. 10, 282–293 (1982)
6. Dubois, D., Prade, H.: A review of fuzzy set aggregation connectives. Inf. Sci. 36, 85–121 (1985)
7. Dubois, D., Prade, H.: Possibility Theory: An Approach to Computerized Processing of Uncertainty. Plenum Press, New York (1988)
8. Dyckhoff, H., Pedrycz, W.: Generalized means as model of compensative connectives. Fuzzy Sets Syst. 14, 143–154 (1984)
9. Grabisch, M.: Fuzzy integrals in multicriteria decision making. Fuzzy Sets Syst. 69, 279–298 (1995)
10. Grabisch, M., Nguyen, H.T., Walker, E.A.: Fundamentals of Uncertainty Calculi With Applications to Fuzzy Inference, B: Mathematical and Statistical Methods, vol. 30. Kluwer Academic Publihers, Dordrecht (1995)
11. Hardy, G.H., Littlewood, J.E., Polya, G.: Inequalities, 2nd edn. Cambridge University Press, London (1973)
12. Kaymak, U.: Fuzzy decision making with control applications. Ph.D. thesis, Delft University of Technology, P.O. Box 5031, 2600 GA, Delft, the Netherlands (1998)
13. Kaymak, U., van Nauta Lemke, H.R.: A sensitivity analysis approach to introducing weight factors into decision functions in fuzzy multicriteria decision making. Fuzzy Sets Syst. 97(2), 169–182 (1998)
14. Kim, K., Park, K.S.: Ranking fuzzy numbers with index of optimism. Fuzzy Sets Syst. 35, 143–150 (1990)
15. Klement, E.P.: Operations on fuzzy setsan axiomatic approach. Inf. Sci. 27(3), 221–232 (1982)
16. Klir, G.J., Folger, T.A.: Fuzzy Sets, Uncertainty and Information. Prentice-Hall, New Jersey (1988)
17. Kovalerchuk, B., Taliansky, V.: Comparison of empirical and computed values of fuzzy conjunction. Fuzzy Sets Syst. 46, 49–53 (1992)
18. Krishnapuram, R., Lee, J.: Fuzzy-connective-based hierarchical aggregation networks for decision making. Fuzzy Sets Syst. 46, 11–27 (1992)
19. Luce, R.D., Raiffa, H.: Games and Decisions. John Wiley, New York (1957)
20. Mizumoto, M.: Pictorial representations of fuzzy connectives, part I: cases of t-norms, t-conorms and averaging operators. Fuzzy Sets Syst. 31, 217–242 (1989)

21. Mizumoto, M.: Pictorial representations of fuzzy connectives, part II: cases of compensatory operators and self-dual operators. Fuzzy Sets Syst. **32**, 45–79 (1989)
22. Ovchinnikov, S.: On robust aggregation procedures. In: Buchon-Meunier, B. (ed.) Aggregation and Fusion of Imperfect Information, Studies in Fuzziness and Soft Computing, vol. 12, pp. 3–10. Springer, Heidelberg (1998)
23. Thole, U., Zimmermann, H.J., Zysno, P.: On the suitability of minimum and product operators for the intersection of fuzzy sets. Fuzzy Sets Syst. **2**, 167–180 (1979)
24. van Nauta Lemke, H.R., Dijkman, J.G., van Haeringen, H., Pleeging, M.: A characteristic optimism factor in fuzzy decision-making. In: Proceedings of IFAC Symposium on Fuzzy Information, Knowledge Representation and Decision Analysis, pp. 283–288. Marseille, France (1983)
25. Yager, R.R.: Fuzzy decision making including unequal objectives. Fuzzy Sets Syst. **1**, 87–95 (1978)
26. Yager, R.R.: A measurement-informational discussion of fuzzy union and intersection. Int. J. Man-Mach. Stud. **11**, 189–200 (1979)
27. Yager, R.R.: On a general class of fuzzy connectives. Fuzzy Sets Sys. **4**, 235–242 (1980)
28. Yager, R.R.: On ordered weighted averaging aggregation operators in multicriteria decision-making. IEEE Trans. Syst. Man Cybern. **18**(1), 183–190 (1988)
29. Zadeh, L.A.: Outline of a new approach to the analysis of complex systems and decision processes. IEEE Trans. Syst. Man Cybern. **3**, 28–44 (1973)
30. Zimmermann, H.J.: Results of empirical studies in fuzzy set theory. In: Klir, G.J. (ed.) Applied General Systems Research: Recent Developments, pp. 303–312. Plenum Press, New York (1978)
31. Zimmermann, H.J.: Fuzzy Sets, Decision Making and Expert Systems. Kluwer Academic Publishers, Boston (1987)
32. Zimmermann, H.J.: Fuzzy Set Theory and its Application, 3rd edn. Kluwer, Boston (1996)
33. Zimmermann, H.J., Zysno, P.: Latent connectives in human decision making. Fuzzy Sets Syst. **4**, 37–51 (1980)

Evolving Possibilistic Fuzzy Modeling and Application in Value-at-Risk Estimation

Leandro Maciel, Rosangela Ballini and Fernando Gomide

Abstract This chapter suggests an evolving possibilistic fuzzy modeling approach for value-at-risk modeling and estimation. The modeling approach is based on an extension of the possibilistic fuzzy c-means clustering and functional fuzzy rule-based systems. It employs memberships and typicalities to update clusters centers and creates new clusters using a statistical control distance-based criteria. Evolving possibilistic fuzzy modeling (ePFM) also uses an utility measure to evaluate the quality of the current cluster structure. The fuzzy rule-based model emerges from the cluster structure. Market risk exposure plays a key role for financial institutions in risk assessment and management. A way to measure risk exposure is to evaluate the losses likely to incur when the prices of the portfolio assets decline. Value-at-risk (VaR) estimate is amongst the most prominent measure of financial downside market risk. Computational experiments are conducted to evaluate ePFM for value-at-risk estimation using data of the main equity market indexes of United States (S&P 500) and Brazil (Ibovespa) from January 2000 to December 2012. Econometric models benchmarks such as GARCH and EWMA, and state of the art evolving approaches are compared against ePFM. The results suggest that ePFM is a potential candidate for VaR modeling and estimation because it achieves higher performance than econometric and alternative evolving approaches.

L. Maciel (✉) · F. Gomide
School of Electrical and Computer Engineering, University of Campinas,
Campinas, São Paulo, Brazil
e-mail: maciel@dca.fee.unicamp.br

F. Gomide
e-mail: gomide@dca.fee.unicamp.br

R. Ballini
Institute of Economics, University of Campinas, Campinas, São Paulo, Brazil
e-mail: ballini@eco.unicamp.br

© Springer International Publishing Switzerland 2017
J. Kacprzyk et al. (eds.), *Granular, Soft and Fuzzy Approaches
for Intelligent Systems*, Studies in Fuzziness and Soft Computing 344,
DOI 10.1007/978-3-319-40314-4_7

1   Introduction

Nowadays Value-at-Risk (VaR) is accepted as one of the most prominent measure of financial downside market risk. One of the major decisions after the Basel Committed on Banking Supervision creation was the use of VaR as a standard mechanism to measure market risk. It recommends commercial banks with significant trade activity to use their own VaR measure to find how much capital they should set aside to cover their market risk exposure, and U.S. bank regulatory agencies to audit the VaR methodology employed by the banks [32, 35].

Despite its theoretical and numerical weakness, such as nonsubaddand non-convexity [14], VaR is the most widely used risk measure in practice. Accurate computation of VaR is fundamental for quantile-based risk measures estimation such as expected shortfall [21]. Although VaR is a relatively simple concept, robust estimation of its values is often neglected. Currently, the generalized autoregressive conditional heteroskedasticity (GARCH) family model [17], a type of non-linear time series model, became a standard approach to estimate the volatility of financial market data [35], the key input for parametric VaR computation.

The literature has shown a growing interest on the use of distinct methods for VaR modeling and estimation. Examples include extreme value theory [40], quantile regression models [17], Bayesian approaches [10], and Markov switching techniques [20]. Kuester et al. [26] gives an overview of these and other models for VaR estimation.

In spite of the recent advances, current econometric models exhibit some limitations. The most important concerns the restrictive assumptions on the distribution of assets returns. There has been accumulated evidence that portfolio returns (or log returns) usually are not normally distributed. In particular, it is frequently found that market returns display structural shifts, negative skewness and excess kurtosis in the distribution of the time series [35].

To address these limitations, recent studies have suggested the use of evolving fuzzy systems (eFS) for volatility modeling and forecasting in VaR estimation [7, 35, 41, 52, 53]. More recent, [39, 46] suggested the use of a cloud-based evolving fuzzy model and a hybrid neural fuzzy network for volatility forecasting. The authors show that evolving fuzzy modeling approaches outperform econometric techniques, and appear as a potential tool to deal with volatility clustering.

Evolving fuzzy systems are an advanced form of adaptive systems because they have the ability to simultaneous learn the model structure and functionality from flows of data. eFS has been useful to develop adaptive fuzzy rule-based models, neural fuzzy, and fuzzy tree models, to mention a few. Examples of the different types of evolving fuzzy rule-based and fuzzy neural modeling approaches include the pioneering evolving Takagi-Sugeno (eTS) modeling [3] approach and extensions (e.g. Simpl_eTS [4], eXtended eTS (xTS) [5]). An autonomous user-free control parameters modeling scheme called eTS+ is given in [2]. The eTS+ uses criteria such as age, utility, local density, and zone of influence to update the model structure.

Later, ePL+ [36] was developed in the realm of participatory learning clustering [50, 51]. ePL+ extends the ePL approach [30] and uses the updating strategy of eTS+.

An alternative method for evolving TS modeling is given in [13] based on a recursive form of the fuzzy c-means (rFCM) algorithm. Clustering in evolving modeling aims at learning the model structure. Later, the rFCM method was translated in a recursive Gustafson-Kessel (rGK) algorithm. Similarly the original off-line GK, the purpose of rGK is to capture different cluster shapes [12]. Combination of the rGK algorithm and evolving mechanisms such as adding, removing, splitting, merging clusters, and recursive least squares became a powerful evolving fuzzy modeling approach called eFuMo [11].

A distinct, but conceptually similar approach for TS modeling is the dynamic evolving neural fuzzy inference system model (DENFIS) [22]. DENFIS uses distance-based recursive clustering to adapt the rule base structure. The weighted recursive least squares with forgetting factor algorithm updates the parameters of rule consequents. A recursive clustering algorithm derived from a modification of the vector quantization technique, called evolving vector quantization, is another effective way to construct flexible fuzzy inference systems (FLEXFIS) [33].

An on-line sequential extreme learning (OS-ELM) algorithm for single hidden layer feedforward neural networks with either additive or radial basis function hidden nodes in a unified framework was developed in [29]. Computational experiments using stream data of benchmark problems drawn from the regression, classification, and time series forecasting has shown that OS-ELM is an efficient tool, with good generalization and computational performance.

An evolving fuzzy modeling approach using tree structures, namely, evolving fuzzy trees (eFT) was introduced in [27]. The eFT model is a fuzzy linear regression tree whose topology can be continuously updated through a statistical model selection test. A fuzzy linear regression tree is a fuzzy tree with a linear model in each leaf. Experiments in the realm of time series forecasting have shown that the fuzzy evolving regression tree is a promising approach for adaptive system modeling.

The fuzzy self-organizing neural network [28] is an alternative recursive modeling scheme based on an error criterion, and on the generalization performance of the network. Similarly, [49] suggests a self-adaptive fuzzy inference network (SaFIN) using a categorical learning-induced partitioning mechanism to cluster data. The key is to avoid the need of prior knowledge on the number of clusters for each dimension of the input-output space. Similar mechanisms to learn model structure and parameters include the self-organizing fuzzy neural network (SOFNN) [43], and the generalized adaptive neural fuzzy inference system (GANFIS) [6]. Other important instance of evolving mechanism includes [45]. A comprehensive source of the evolving approaches can be found in [34].

Recently [37] developed a recursive possibilistic fuzzy modeling (rPFM) approach whose purpose is to improve model robustness against noisy data and outliers. The ePFM uses a recursive form of the possibilistic fuzzy c-means (PFCM) clustering algorithm suggested by [42] to develop TS fuzzy rule-based model structure, and employs the weighted recursive least squares algorithm to estimate the parameters of affine functions of the rule consequents. The PFCM simultaneously produces memberships and typicalities to alleviate outliers and noisy data sensitivity of traditional fuzzy clustering approaches, yet avoids coincident clusters [25, 42]. The advantages of PFCM have been emphasized in the literature [9, 18, 48]. Using traditional benchmarks of time series forecasting problems, [37] showed the potential of rPFM to deal with nonlinear and nonstationary systems whose data affected by noise and outliers.

Despite the potential of the rPFM modeling approach to handle noisy data and outliers, it assumes that the model structure, i.e. the number of clusters or, equivalently, the number of TS fuzzy rules, is defined by the user. This is a major limitation in adaptive system modeling, especially when handling nonstationary data. To overcome this limitation, [38] suggested an evolving possibilistic fuzzy modeling (ePFM) for data streams. ePFM adapts its structure with a possibilistic extension of the evolving Gustafson-Kessel-Like algorithm (eGKL) suggested in [19]. Like [19], creation of new clusters is determined by a statistical control distance-based criteria, but cluster structure update uses both memberships and typicalities. The model incorporates the advantages of the GK clustering algorithm by identifying clusters with different shape and orientation while processing data streams. ePFM also uses an utility measure to evaluate the quality of the current cluster structure. The utility measure allows the rule base to shrink by removing rules with low utility (the data pattern shifted away from the domain of the rule) and gives a simpler and more relevant rule base to encapsulate the current state of the process as mirrored by recent data.

Further, the aim of the chapter is to address evolving possibilistic fuzzy modeling and value-at-risk estimation. The possibilistic approach is important because financial markets are often affected by news, expectations, and investors psychological states, which induce volatility, noisy information, and cause outliers. In this situation, possibilistic modeling has the potential to attenuate the effect of noise and outliers when building financial volatility forecasting models. Moreover, ePFM is able to handle nonlinear and time-varying dynamics such as assets returns volatility using data streams, which is essential for real-time decision making in risk management. Computational experiments were performed to compare ePFM against current econometric benchmarks and alternative state of the art evolving fuzzy and neurofuzzy models for VaR estimation using daily data from January 2000 to December 2012 of the main equity market indexes in the United States (S&P 500) and Brazil (Ibovespa).

After this introduction, the chapter proceeds as follows. Section 2 briefly recalls TS modeling and possibilistic fuzzy c-means clustering. Next, Sect. 3 details the evolving possibilistic fuzzy modeling approach. Section 4 evaluates the performance and compares ePFM against GARCH, EWMA, and evolving approaches such as eTS+, ePL+, DENFIS, eFuMo, and OS-ELM. Section 5 concludes the chapter and lists issues for further investigation.

2   TS Model and Possibilistic Fuzzy C-means

This section briefly describes the basic constructs of Takagi-Sugeno modeling and its identification tasks, as well as the possibilistic fuzzy c-means clustering algorithm.

2.1   Takagi-Sugeno Fuzzy Model

Takagi-Sugeno (TS) fuzzy model with affine consequents consists of a set of fuzzy functional rules of the following form:

$$\mathcal{R}_i : \text{ IF } \mathbf{x} \text{ is } \mathcal{A}_i \text{ THEN } y_i = \theta_{i0} + \theta_{i1}x_1 + \cdots + \theta_{im}x_m, \tag{1}$$

where $\mathcal{R}_i$ is the ith fuzzy rule, $i = 1, 2, \ldots, c$, c is the number of fuzzy rules, $\mathbf{x} = [x_1, x_2, \ldots, x_m]^T \in \mathfrak{R}^m$ is the input, $\mathcal{A}_i$ is the fuzzy set of the antecedent of the ith fuzzy rule and its membership function $\mathcal{A}_i(\mathbf{x}) : \mathfrak{R}^m \rightarrow [0, 1]$, $y_i \in \mathfrak{R}$ is the output of the ith rule, and θ_{i0} and $\theta_{ij}, j = 1, \ldots, m$, are the parameters of the consequent of the ith rule.

Fuzzy inference using TS rules (1) has a closed form as follows:

$$y = \sum_{i=1}^{c} \left(\frac{\mathcal{A}_i(\mathbf{x})y_i}{\sum_{j=1}^{c} \mathcal{A}_j(\mathbf{x})} \right). \tag{2}$$

The expression (2) can be rewritten using normalized degree of activation:

$$y = \sum_{i=1}^{c} \lambda_i y_i = \sum_{i=1}^{c} \lambda_i \mathbf{x}_e^T \theta_i, \tag{3}$$

where

$$\lambda_i = \frac{\mathcal{A}_i(\mathbf{x})}{\sum_{j=1}^{c} \mathcal{A}_j(\mathbf{x})}, \tag{4}$$

is the normalized firing level of the ith rule, $\theta_i = [\theta_{i0}, \theta_{i1}, \ldots, \theta_{im}]^T$ is the vector of parameters, and $\mathbf{x}_e = [1 \ \mathbf{x}^T]^T$ is the expanded input vector.

The TS model uses parametrized fuzzy regions and associates each region with an affine (local) model. The non-linear nature of the rule-based model emerges from the fuzzy weighted combination of the collection of the multiple local affine models. The contribution of a local model to the model output is proportional to the degree of firing of each rule.

TS modeling requires to learn the antecedent part of the model using e.g. a fuzzy clustering algorithm, and to estimate the parameters of the affine consequents.

2.2 Possibilistic Fuzzy C-means Clustering

The possibilistic fuzzy c-means clustering algorithm [42] can be summarized as follows. Let $\mathbf{x}_k = [x_{1k}, x_{2k}, \ldots, x_{mk}]^T \in \mathfrak{R}^m$ be the input data at k. A set of n inputs is denoted by $X = \{\mathbf{x}_k, k = 1, \ldots, n\}$, $X \subset \mathfrak{R}^{m \times n}$. The aim of clustering is to partition the data set X into c subsets (clusters).

A possibilistic fuzzy partition of the set X is a family $\{\mathscr{A}_i, 1 \leq i \leq c\}$. Each $\mathscr{A}_i$ is characterized by membership degrees and typicalities specified by the fuzzy and typicality partition matrices $U = [u_{ik}] \in \mathfrak{R}^{c \times n}$ and $T = [t_{ik}] \in \mathfrak{R}^{c \times n}$, respectively. The entries of the ith row of matrix U (T) are the values of membership (typicalities) degrees of the data $\mathbf{x}_1, \ldots, \mathbf{x}_n$ in $\mathscr{A}_i$.

The possibilistic fuzzy c-means (PFCM) clustering algorithm produces c vectors that represent c cluster centers. The PFCM algorithm derives from the solution of the following optimization problem:

$$\min_{U,T,V} \left\{ J = \sum_{k=1}^{n} \sum_{i=1}^{c} (au_{ik}^{\eta_f} + bt_{ik}^{\eta_p}) D_{ik}^2 + \sum_{i=1}^{c} \gamma_i \sum_{k=1}^{n} (1 - t_{ik})^{\eta_p} \right\}, \tag{5}$$

subject to

$$\sum_{i=1}^{c} u_{ik} = 1 \; \forall \, k,$$

$$0 \leq u_{ik}, t_{ik} \leq 1 \tag{6}$$

Here $a > 0$, $b > 0$, and $\eta_f > 1$, $\eta_p > 1$, $\gamma_i > 0$ are user defined parameters, and D_{ik}^2 is the distance of $\mathbf{x}_k$ to the ith cluster centroid $\mathbf{v}_i$. The constants a and b define the relative importance of fuzzy membership and typicality values in the objective function, respectively. $V = [\mathbf{v}_1, \mathbf{v}_2, \ldots, \mathbf{v}_c]^T \in \mathfrak{R}^{c \times m}$ is the matrix of cluster centers, η_f and η_p are parameters associated with membership degrees and typicalities, respectively, with default value $\eta_f = \eta_p = 2$.

If $D_{ik}^2 > 0$ for all i, and X contains at least c distinct data points, then $(U, T, V) \in M_f \times M_p \times \mathfrak{R}^{c \times n}$ minimizes J, with $1 \leq i \leq c$ and $1 \leq k \leq n$, only if [42]:

$$u_{ik} = \left(\sum_{j=1}^{c} \left(\frac{D_{ik}}{D_{jk}} \right)^{2/(\eta_f - 1)} \right)^{-1}, \tag{7}$$

$$t_{ik} = \frac{1}{1 + \left(\frac{b}{\gamma_i} D_{ik}^2 \right)^{1/(\eta_p - 1)}}, \tag{8}$$

$$\mathbf{v}_i = \frac{\sum_{k=1}^{n} (au_{ik}^{\eta_f} + bt_{ik}^{\eta_p}) \mathbf{x}_k}{\sum_{k=1}^{n} (au_{ik}^{\eta_f} + bt_{ik}^{\eta_p})}, \tag{9}$$

where

$$M_p = \left\{ T \in \mathfrak{R}^{c \times n} : 0 \leq t_{ik} \leq 1, \forall\, i, k; \forall\, k\, \exists\, i \ni t_{ik} > 0 \right\}, \tag{10}$$

$$M_f = \left\{ U \in M_p : \sum_{i=1}^{c} u_{ik} = 1 \,\forall\, k; \sum_{k=1}^{n} u_{ik} > 0 \,\forall\, i \right\}, \tag{11}$$

are the sets of possibilistic and fuzzy partition matrices, respectively.

Originally, [42] recommends to choose parameters γ_i as follows:

$$\gamma_i = K \frac{\sum_{k=1}^{n} u_{ik}^{\eta_f} D_{ik}^2}{\sum_{k=1}^{n} u_{ik}^{\eta_f}}, \quad 1 \leq i \leq c, \tag{12}$$

where $K > 0$ (usually $K = 1$), and u_{ik} are entries of a terminal FCM partition of X.

3 Evolving Possibilistic Fuzzy Modeling

The evolving possibilistic fuzzy modeling approach (ePFM), suggested by [38], extends the evolving Gustafson-Kessel-Like clustering algorithm (eGKL), suggested in [19]. ePFM considers both membership and typicalities to update the cluster structure, i.e. the antecedents of TS fuzzy rule-based model, and incorporates a utility measure to avoid unused clusters.

3.1 Antecedents Identification

ePFM proceeds based on the underlying objective function for possibilistic fuzzy clustering algorithm as in (5). In such case, the distance D_{ik} is the same as used by the Gustafson-Kessel algorithm, i.e. the Mahalanobis distance, which is a squared inner-product distance norm that depends on a positive definite symmetric matrix A_{ik} as follows:

$$D_{ik}^2 = ||\mathbf{x}_k - \mathbf{v}_i||_{A_{ik}}^2 = (\mathbf{x}_k - \mathbf{v}_i)^T A_{ik} (\mathbf{x}_k - \mathbf{v}_i). \tag{13}$$

The matrix A_{ik}, $i = 1, \ldots, c$, determines the shape and orientation of the cluster i, i.e., it is an adaptive norm unique for every cluster, calculated by estimates of the data dispersion:

$$A_{ik} = \left[\rho_i \det(F_{ik}) \right]^{1/m} F_{ik}^{-1}, \tag{14}$$

where ρ_i is the cluster volume of the ith cluster (usually $\rho_i = 1$ for all clusters) and F_{ik} is the fuzzy dispersion matrix:

$$F_{ik} = \frac{\sum_{k=1}^{n} u_{ik}^{\eta_f} \left(\mathbf{x}_k - \mathbf{v}_i\right) \left(\mathbf{x}_k - \mathbf{v}_i\right)^T}{\sum_{k=1}^{n} u_{ik}^{\eta_f}}. \tag{15}$$

Most of the fuzzy clustering algorithms assume clusters with spherical shapes. Actually, in real world applications clusters often have different shapes and orientations in the data space. A way to distinguish cluster shapes is to use information about the dispersion of the input data as the Mahalanobis distance does.

The antecedents identification of ePFM extends possibilistic fuzzy clustering algorithm using Mahalanobis distance to deal with streams of data. The evolving mechanisms, i.e. creation and update of clusters, are based on the evolving Gustafson-Kessel-Like clustering algorithm principles [19], which are inspired by two common recursive clustering algorithm: the k-nearest neighbor (k-NN) [23] and the linear vector quantization (LVQ) [24].

Suppose $\mathbf{x}_k$ is a input data at step k. Two possibilities should be considered to update the current cluster structure. First, the data may belong to an existing cluster, within the cluster boundary, which requires just a cluster update. Otherwise, it may define a new cluster. These scenarios are considered in detail in what follows.

Suppose that we have c clusters when the kth data is input. The similarity between the new data $\mathbf{x}_k$ and each of the existing c clusters is evaluated using the Mahalanobis distance (13). The similarity relation is evaluated by checking the following condition:

$$D_{ik}^2 < \chi_{m,\beta}^2, \quad i = 1, \ldots, c, \tag{16}$$

where $\chi_{m,\beta}^2$ is the $(1 - \beta)$th value of the chi-squared distribution with m degrees of freedom and β is the probability of false alarm.

Relation (16) comes from statistical process control to identify variations in systems that are due to actual input changes rather than process noise. For details on criteria (16) see [19].

If the condition (16) holds, then the process is under control. The minimal distance D_{ik} determines the closest cluster p as

$$p = \arg \min_{i=1,\ldots,c} (D_{ik}), \quad D_{ik}^2 < \chi_{m,\beta}^2, \quad i = 1, \ldots, c. \tag{17}$$

In this case, the vector $\mathbf{x}_k$ is assigned to the pth cluster and

$$M_{p,new} = M_{p,old} + 1, \tag{18}$$

where M_i counts the number of data points that fall within the boundary of cluster i, $i = 1, \ldots, c$.

As [19] suggests for the eGKL algorithm, the pth cluster center could be updated using the Kohonen-like rule [24]:

$$\mathbf{v}_{p,new} = \mathbf{v}_{p,old} + \alpha(\mathbf{x}_k - \mathbf{v}_{p,old}), \tag{19}$$

where α is a learning rate, $\mathbf{v}_{p,new}$ and $\mathbf{v}_{p,old}$ denote the new and old values of the cluster center. Notice that $\alpha(\mathbf{x}_k - \mathbf{v}_{p,old})$ may be viewed as a term proportional to the gradient of an Euclidean distance based objective function, such as in FCM. The evolving possibilistic fuzzy modeling suggested in this chapter updates the pth cluster center taking into account a term proportional the gradient of the possibilistic fuzzy clustering objective function in (5). Thus, the pth cluster center is updated as follows:

$$\mathbf{v}_{p,new} = \mathbf{v}_{p,old} + \alpha(au_{pk}^{n_f} + bt_{pk}^{n_p})A_{pk}(\mathbf{x}_k - \mathbf{v}_{p,old}). \tag{20}$$

The determinant and the inverse of the dispersion matrix of the pth cluster is updated as follows[1]:

$$F_{p,new}^{-1} = (I - G_p(\mathbf{x}_k - \mathbf{v}_{p,old}))F_{p,old}^{-1}\frac{1}{1-\alpha}, \tag{21}$$

$$\det(F_{p,new}) = (1-\alpha)^{m-1}\det(F_{p,old})(1 - \alpha + \alpha(\mathbf{x}_k - \mathbf{v}_{p,old})^T F_{p,old}^{-1}(\mathbf{x}_k - \mathbf{v}_{p,old}), \tag{22}$$

where

$$G_p = F_{p,old}^{-1}(\mathbf{x}_k - \mathbf{v}_{p,old})\frac{\alpha}{1 - \alpha + \alpha(\mathbf{x}_k - \mathbf{v}_{p,old})^T F_{p,old}^{-1}(\mathbf{x}_k - \mathbf{v}_{p,old})}, \tag{23}$$

and I is an identity matrix of order m.

Simultaneously the remaining clusters centers are updated in the opposite direction to move them away from the pth cluster:

$$\mathbf{v}_{q,new} = \mathbf{v}_{q,old} - \alpha(au_{qk}^{n_f} + bt_{qk}^{n_p})A_{qk}(\mathbf{x}_k - \mathbf{v}_{q,old}), \quad q = 1, \ldots, c, \ q \neq p. \tag{24}$$

The parameter M_i, $i = 1, \ldots, c$, assesses the credibility of the estimated clusters. According to [19], its minimal value M_{min} corresponds to the minimal number of data points needed to learn the parameters of the ith inverse dispersion matrix F_{ik}^{-1}, that is estimated by the dimension m of the data vector:

$$M_{min} = Qm(m + 1)/2, \tag{25}$$

where Q is the credibility parameter, with default value $Q = 2$.

---

[1]The computation details are found in [19].

On the other hand, if (16) does not hold, then $\mathbf{x}_k$ is not similar to any of the cluster centers. Thus, the natural action is to create a new cluster. However, one must check whether that fact is not due to the lack of credible clusters surrounding $\mathbf{x}_k$, that is, whether the condition:

$$M_p < M_{min}, \tag{26}$$

applies, $p = \arg\min_{i=1,\ldots,c}(D_{ik})$. If this is the case, then the closest cluster p is updated using (20)–(23). Otherwise, a new cluster is created, $c_{new} = c_{old} + 1$, with the following initialization:

$$\mathbf{v}_{c,new} = \mathbf{x}_k, \; F_{c,new}^{-1} = F_0^{-1} = \kappa I, \; \det(F_{c,new}) = \det(F_0), \; M_{c,new} = 1, \tag{27}$$

where I is an identity matrix of size m and κ is a sufficient large positive number.

The initialization in (27) is also used if there is no initially collected data set, supposing that $\mathbf{x}_1$ represents the very first data point of the data streams $\mathbf{x}_k$, $k = 1, 2, \ldots$.

3.2   Cluster Quality Measurement

In this paper, the quality of the cluster structure is monitored at each step considering the utility measure introduced in [2]. The utility measure is an indicator of the accumulated relative firing level of a corresponding rule:

$$\mathcal{U}_{ik} = \frac{\sum_{l=1}^{k} \lambda_i}{k - K^{i*}}, \tag{28}$$

where K^{i*} is the step that indicates when cluster $i*$ was created.

Once a rule is created, the utility indicates how much the rule has been used. This quality measure aims at avoiding unused clusters kept in the structure. Clusters corresponding with low quality fuzzy rules can be deleted. Originally, [2] suggested, at each step k, the following criteria: if $\mathcal{U}_{ik}$ is less or equal than a threshold, specified by the user, then the ith cluster is removed. To turn the ePFM algorithm more autonomous, the following criteria to remove low quality clusters is suggested:

$$\text{If } \mathcal{U}_{ik} \leq (\bar{\mathcal{U}}_i - 2\sigma_{\mathcal{U}_i}) \text{ Then } c_{new} = c_{old} - 1, \tag{29}$$

where $\bar{\mathcal{U}}_i$ and $\sigma_{\mathcal{U}_i}$ are the sample average and the standard deviation of the utility of cluster i values.

This condition means that if the utility of cluster i at k is less or equal than 2 standard deviation of the average utility of cluster i, then cluster i has low utility and it is removed. This idea relates to the 2σ process control band in statistical process control but considering the tail of the left side of utility distribution, since it is associated with the most unused clusters (lower utility). This principle guarantees high relevance cluster structure and corresponding fuzzy local models. Alternative quality measures such as age, support, zone of influence and local density may be adopted.

3.3 Consequent Parameters Estimation

Estimation of the parameters of the affine rule consequents is done using weighted recursive least squares algorithm (wRLS) [31] as in [2]. Expression (3) can be rewritten as:

$$y = \Lambda^T \Theta, \tag{30}$$

where $\Lambda = \left[\lambda_1 \mathbf{x}_e^T, \lambda_2 \mathbf{x}_e^T, \ldots, \lambda_c \mathbf{x}_e^T\right]^T$ is the fuzzily weighted extended input, $\mathbf{x}_e = \left[1 \ \mathbf{x}^T\right]^T$ the expanded data vector, and $\Theta = \left[\theta_1^T, \theta_2^T, \ldots, \theta_c^T\right]^T$ the parameter matrix, $\theta_i = [\theta_{i0}, \theta_{i1}, \ldots, \theta_{im}]^T$.

Given that the actual output can be obtained at each step, the parameters of the consequents can be updated using the recursive least squares (RLS) algorithm considering local or global optimization. In this chapter we use the locally optimal error criterion wRLS:

$$\min_{\theta_i} E_L^i = \min_{\theta_i} \sum_{k=1}^{n} \lambda_i \left(y_k - \mathbf{x}_{ek}^T \theta_i\right)^2. \tag{31}$$

Therefore, parameters of the rule consequents are updated as follows [2, 31]:

$$\theta_{i,k+1} = \theta_{ik} + \Sigma_{ik} \mathbf{x}_{ek} \lambda_{ik} \left(y_k - \mathbf{x}_{ek}^T \theta_{ik}\right), \ \theta_{i0} = 0, \tag{32}$$

$$\Sigma_{i,k+1} = \Sigma_{ik} - \frac{\lambda_{ik} \Sigma_{ik} \mathbf{x}_{ek} \mathbf{x}_{ek}^T \Sigma_{ik}}{1 + \lambda_{ik} \mathbf{x}_{ek}^T \Sigma_{ik} \mathbf{x}_{ek}}, \ \Sigma_{i0} = \Omega I, \tag{33}$$

where I is an identity matrix of size m, Ω a large number (usually $\Omega = 1000$), and $\Sigma \in \mathfrak{R}^{m \times m}$ the dispersion matrix.

3.4 EPFM Algorithm

The evolving possibilistic fuzzy modeling (ePFM) approach is summarized next. The steps of the algorithm are non-iterative. The procedure adapts an existing model whenever the pattern encoded in data changes. Its recursive nature means that, as far as data storage is concerned, it is memory efficient.

Evolving possibilistic fuzzy modeling

1. Compute γ_i and $\mathbf{v}_{i0}$ as terminal FCM partition, and $M_{min} = Qm(m+1)/2$

2. Choose control parameters a, b, α, β, κ, and initialize $F_0^{-1} = \kappa I$

3. for $k = 1, 2, \dots$ do

4.   read the next data $\mathbf{x}_k$

5.   check the similarity of $\mathbf{x}_k$ to existing clusters: $D_{ik}^2 < \chi_{m,\beta}^2$, $i = 1, \dots, c$

6.   identify the closest cluster: $p = \arg\min_{i=1,\dots,c}(D_{ik})$

7.   if $D_{ik}^2 < \chi_{m,\beta}^2$ or $M_p < M_{min}$ then:

8.   update the parameters of the pth cluster using (20)–(23)

9.   move away the centres of remaining cluster using (24)

10.   else

11.   create a new cluster: $c_{new} = c_{old} + 1$

12.   initilialize the new cluster using (27)

13.   end if

14.   if $\mathcal{U}_{ik} \leq (\bar{\mathcal{U}}_i - 2\sigma_{\mathcal{U}_i})$ then delete cluster i:

15. $c_{new} = c_{old} - 1$

16.   end if

17.   compute rule consequent parameters using the wRLS

18.   compute model output y_{k+1}

19. end for

4 Computational Experiments

The ePFM approach introduced in this chapter gives a flexible modeling procedure and can be applied to a range of problems such as process modeling, time series forecasting, classification, system control, and novelty detection. This section evaluates the performance of ePFM for Value-at-Risk modeling and estimation of S&P 500 and Ibovespa indexes in terms of volatility forecasting. The results of ePFM are compared with GARCH and EWMA models, and with state of the art evolving fuzzy, neuro and neuro-fuzzy modeling approaches, the eTS+ [2], ePL+ [36], rFCM [13], DENFIS [22], eFuMo [11], and OLS-ELM [29].

4.1 VaR Estimation

Value-at-Risk (VaR) has been adopted by practitioners and regulators as the standard mechanism to measure market risk of financial assets. It encapsulates in a single quantity the potential market value loss of a financial asset over a time horizon h, at a significance or coverage level α_{VaR}. Alternatively, it reflects the asset market value loss over the time horizon h, that is not expected to be exceeded with probability $(1 - \alpha_{VaR})$, i.e.:

$$\Pr\left(r_{k+h} \leq \text{VaR}_{k+h}^{\alpha_{VaR}}\right) = 1 - \alpha_{VaR} \tag{34}$$

where

$$r_{k+h} = \frac{ln(P_{k+h})}{ln(P_k)}, \tag{35}$$

is the asset log return over the period h and P_j is the asset price at j.

Hence, VaR is the α_{VaR}th quantile of the conditional returns distribution defined as: $\text{VaR}_{\alpha_{VaR}} = CDF_{k+h}^{-1}(\alpha_{VaR})$, where $CDF(\cdot)$ is the returns cumulative distribution function and $CDF^{-1}(\cdot)$ denotes its inverse. Here, we concentrate at $h = 1$ as it bears the greatest practical interest.

Let us assume that the daily conditional heteroskedastic returns in (35) of a financial asset can be described by the following process:

$$r_k = \sigma_k z_k, \tag{36}$$

where $z_k \sim i.i.d(0, 1)$ and σ_k is the asset volatility at k.

Therefore, the VaR at $k + 1$ is given by:

$$\text{VaR}_{k+1}^{\alpha_{VaR}} = \sigma_{k+1} CDF_z^{-1}(\alpha_{VaR}), \tag{37}$$

where $CDF_z^{-1}(\alpha_{VaR})$ is the critical value from the normal distribution table at α_{VaR} confidence level. In this section $\alpha_{VaR} = 5\%$ confidence level is assumed for all models.

In a VaR forecasting context, volatility modeling plays a crucial role and thus it should place emphasis on the volatility models implemented. This work chooses to model the conditional variance of the returns process with two econometric volatility models and some evolving methods.

4.2 Econometric Benchmarks

Two benchmark econometric approaches, GARCH and EWMA, are adopted to estimate the VaR given a α_{VaR} confidence level in (37). The GARCH(r, s) model is as follows [16]:

$$\sigma_k^2 = \delta_0 + \sum_{j=1}^{r} \delta_j r_{k-j}^2 + \sum_{l=1}^{s} \mu_l \sigma_{k-l}^2, \tag{38}$$

where $\delta_j, j = 0, 1, \ldots, r$, and $\mu_l, l = 1, 2, \ldots, s$, are model parameters.

The exponentially weighted moving average (EWMA) model of Riskmetrics has the following form [44]:

$$\sigma^2_{k+1} = \lambda^E \sigma^2_k + (1 - \lambda^E) r^2_k, \tag{39}$$

where λ^E is the forgetting factor with default value $\lambda^E = 0.94$.

The evolving models use the following representation to estimate VaR:

$$\sigma^2_{k+1} \cong r^2_{k+1} = f(r^2_k, r^2_{k-1}, \ldots, r^2_{k-p}), \tag{40}$$

where p is the number of lags considered as input of the evolving models. The number of lags is chosen looking at the partial autocorrelation function of the squared indexes returns.

4.3 Performance Evaluation

The performance of the models is evaluated using two loss functions: the violation ratio and the average square magnitude function. The violation ratio (VR) is the percentage occurrence of an actual loss greater than the estimated maximum loss in the VaR framework. VR is computed as follows:

$$VR = \frac{1}{n} \sum_{k=1}^{n} \Phi_k, \tag{41}$$

where $\Phi_k = 1$ if $r_k < VaR_k$ and $\Phi_k = 0$ if $r_k \geq VaR_k$, where VaR_k is the one step ahead forecasted VaR for day k, and n is the number of observations in the test set.

The average square magnitude function (ASMF) [15] considers the amount of possible default measuring the average squared cost of exceptions. It is computed using:

$$ASMF = \frac{1}{\vartheta} \sum_{j=1}^{\vartheta} \xi_j, \tag{42}$$

where ϑ is the number of exceptions of the respective model, $\xi_j = (r_j - VaR_j)^2$ when $r_j < VaR_j$ and $\xi_j = 0$ when $r_j \geq VaR_j$. The average squared magnitude function enables us to distinguish between models with similar or identical hit rates [35].

All modeling approaches are also characterized in terms of the average number of rules/nodes and the (CPU) time needed to process test data. All algorithms were implemented and run using Matlab® on a laptop equipped with 4 GB and Intel®i3CPU.

Table 1 Statistics of S&P 500 and Ibovespa index returns

Statistic	S&P 500	Ibovespa
Mean	0.0002	0.0004
Std. dev.	0.0136	0.0191
Skewness	−0.1578	−0.1177
Kurtosis	10.2670	6.7847
Max.	0.1096	0.1368
Min.	−0.0947	−0.1210
JB	7094.1	1928.0
p-value	0.0010	0.0010

4.4 Data

The computational results were produced using daily values of the S&P 500 and Ibovespa indexes from January 2000 to December 2012.[2] This period was selected because both equity markets were under stable movements and volatile dynamics due to the international crisis that occurred in that period. The Ibovespa index illustrates how the models perform in emergent economies like the Brazilian. The data was split in two sets. The training set includes the period from January 2000 to December 2003. The remaining data comprises the test set.

4.5 Results

Summary statistics of S&P 500 and Ibovespa index returns are presented in Table 1. For both index returns series the mean are close to zero, and the Ibovespa index has the higher standard deviation. The series also presented a negative skewness, which indicates left a side fat tail, a stylized fact of financial assets returns. The series also show high kurtosis coefficients. The Jarque-Bera statistics (JB) [8] reveal that the return series are non-normal with a 99 % confidence interval. Figure 1 shows the returns series of S&P 500 and Ibovespa. In both markets the volatility clusters are presented, mainly in cases of more unstable periods, such as during the recent crisis that started in 2008 in the USA and then affected other economies like the Brazilian. Moreover, the returns of the Brazilian equity market index are more volatile than the S&P 500, which is expected in emergent economies.

The parameters r and s of the GARCH(r, s) model were selected according to the Bayesian Information Criteria (BIC) [47]. The final GARCH models, GARCH(1, 2) and GARCH(1, 1) for S&P 500 and Ibovespa index returns, respectively, are:

---

[2]The data was provided by Bloomberg.

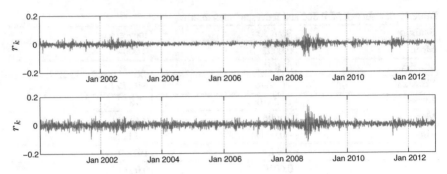

Fig. 1 Daily S&P 500 (*top*) and Ibovespa (*bottom*) index returns from January 2000 to December 2012

$$\sigma_k^2 = 0.0216r_{k-1}^2 + 0.0029\sigma_{k-1}^2 + 0.1031\sigma_{k-2}^2, \tag{43}$$

$$\sigma_k^2 = 0.5029r_{k-1}^2 + 0.4230\sigma_{k-1}^2. \tag{44}$$

The GARCH models indicate that Ibovespa returns volatility is more persistent than S&P 500 due to its estimated coefficients ($\delta_1 + \mu_1 = 0.5029 + 0.4230 = 0.9592$ for Ibovespa and $\delta_1 + \mu_1 + \mu_2 = 0.0216 + 0.0029 + 0.1031 = 0.1276$ for S&P 500), which is in line with the literature for volatility behavior in emergent economies [1].

For the evolving methods, the analysis of the partial autocorrelation function of the squared index returns indicated $p = 2$ and $p = 4$ for S&P 500 and Ibovespa, respectively:

$$\sigma_{k+1}^2 \cong r_{k+1}^2 = f(r_k^2, r_{k-1}^2, r_{k-2}^2), \tag{45}$$

for S&P 500 and

$$\sigma_{k+1}^2 \cong r_{k+1}^2 = f(r_k^2, r_{k-1}^2, r_{k-2}^2, r_{k-3}^2, r_{k-4}^2), \tag{46}$$

for Ibovespa.

The control parameters of ePFM were chosen based on experiments conducted to find the best performance in terms of VR and ASMF measures. ePFM modeling uses $a = 2$, $b = 3$, $\alpha = 0.08$ and $\kappa = 50$ for S&P 500; and $a = 1$, $b = 2$, $\alpha = 0.13$ and $\kappa = 50$ for Ibovespa. Initialization uses the FCM algorithm to choose γ_i and $\mathbf{v}_{i0}$. One must note that is necessary to choose the probability of a false alarm β and to define $\chi_{m,\beta}^2$. This work considers a default probability of false alarm $\beta = 0.0455$ that relates to the 2σ process control band in the single-variable statistical process control as [19]. Table 2 show the $\chi_{m,0.0455}^2$ for different values of m. Therefore, ePFM uses $\chi_{3,0.0455}^2 = 8.0249$ for S&P 500 and $\chi_{5,0.0455}^2 = 11.3139$ for Ibovespa. Control parameters of the alternative evolving methods were also chosen based on simulations to find the best VR and ASMF values.

Table 2 $\chi^2_{m,0.0455}$ for different values of m

m	2	3	4	5	6	7	8	9	10
$\chi^2_{m,0.0455}$	6.1801	8.0249	9.7156	11.3139	12.8489	14.3371	15.7891	17.2118	18.6104

Table 3 Performance evaluation for S&P 500 index VaR estimation

Method	VR (%)	ASMF (%)	# rules (aver.)	Time (s)
GARCH	3.635	0.165	–	44.87
EWMA	3.598	0.160	–	5.211
eTS+	1.814	0.089	3.870	1.774
ePL+	1.992	0.091	4.192	1.801
rFCM	2.075	0.107	4.227	1.967
DENFIS	2.980	0.180	11	18.92
eFuMo	1.712	0.069	3.711	1.700
OS-ELM	2.862	0.173	9	15.30
ePFM	1.690	0.067	3.908	1.882

Table 4 Performance evaluation for Ibovespa index VaR estimation

Method	VR (%)	ASMF (%)	# rules (aver.)	Time (s)
GARCH	4.367	0.214	–	46.34
EWMA	4.567	0.279	–	7.019
eTS+	1.982	0.106	4.871	1.876
ePL+	2.014	0.117	6.740	1.994
rFCM	2.276	0.098	5.306	2.035
DENFIS	3.009	0.192	15	21.48
eFuMo	1.874	0.079	4.652	1.715
OS-ELM	2.933	0.164	13	16.30
ePFM	1.667	0.085	4.891	1.894

Tables 3 and 4 show the violation ratio VR and average square magnitude function ASMF values of the ePFM against the remaining approaches for S&P 500 and Ibovespa VaR estimation using test data, respectively. The results are similar for both indexes. The ePFM model achieves competitive results in terms of VR and ASMF when compared against eFuMo and remaining evolving models for S&P 500 and Ibovespa VaR estimation. The GARCH and EWMA models achieve the worst performance, with higher values of violation ratio and average squared magnitude function for both indexes. For S&P 500 VaR estimation, the evolving possibilistic fuzzy approach reduces the VR and ASMF values in approximately 53.27 % and 58.77 %, respectively, when compared against the econometric benchmark models, GARCH and EWMA (Table 3). Similarly, ePFM reduces the VR and ASMF values in approx-

imately 62.17 % and 72.82 %, respectively, when compared against GARCH and
EWMA (Table 4) for Ibovespa VaR estimation. The gain of performance of using
ePFM modeling approach is even higher when Ibovespa VaR estimation is considered. These results are in line with [39, 46] which suggest that clustering based techniques perform better for volatility modeling and forecasting by handling volatility
clustering, since in this empirical study eTS+, ePL+, rFCM, eFuMo and ePFM show
the lowest values of VR and ASMF against the other models.

Except for GARCH, EWMA, DENFIS and OS-ELM, the computational performance of the evolving approaches are similar in terms of CPU time, considering all
test data processing. eFuMo develops the smallest average number of rules among
all approaches. A smaller number of rules reduces model complexity and enhances
interpretability. Generally speaking, all evolving approaches are qualified to deal
with on-line stream data processing for VaR estimation in risk management decision
making.

Figures 2 and 3 show the return and VaR estimates produced by ePFM modeling
approach for S&P 500 and Ibovespa, respectively. Notice the high adequacy of ePFM
to capture volatility dynamics in terms of VaR estimates in both economies.

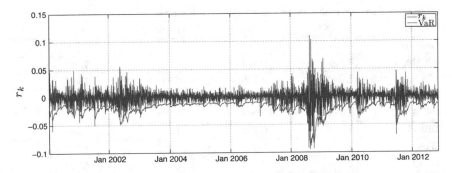

Fig. 2   Daily S&P 500 returns and VaR estimates using ePFM

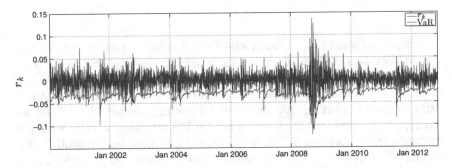

Fig. 3   Daily Ibovespa returns and VaR estimates using ePFM

5 Conclusion

This chapter has suggested an evolving possibilisitc fuzzy modeling approach for Value-at-Risk (VaR) estimation. The approach combines recursive possibilistic fuzzy clustering to learn the model structure, and a weighted recursive least squares to estimate the model parameters. The idea is to adapt the model structure and parameter whenever required by new input data. ePFM creates new clusters using a statistical control distance-based criteria, and clusters are updated using memberships and typicalities. The model incorporates the advantages of the Gustafson-Kessel clustering algorithm of identifying clusters with different shape and orientation while processing data streams. ePFM also uses an utility measure to evaluate the quality of the current cluster/model structure. Computational experiments addressed ePFM, econometric GARCH and EWMA benchmarks, and state of the art evolving fuzzy and neuro-fuzzy models for VaR estimation using daily data from January 2000 to December 2012 of the main equity market indexes in the United States (S&P 500) and Brazil (Ibovespa). Results indicate the superior performance of ePFM and all the evolving techniques against GARCH and EWMA benchmarks in both economies considered. Further work shall generalize ePFM to handle mixture of cluster shapes, to make the ePFM algorithm fully autonomous, and to evaluate the ePFM model in risk management strategies.

Acknowledgments The authors thank the Brazilian Ministry of Education (CAPES), the Brazilian National Research Council (CNPq) grant 304596/2009-4, and the Research of Foundation of the State of São Paulo (FAPESP) for their support.

References

1. Aggarwal, R., Inclan, C., Leal, R.: Volatility in emergent stock markets. J. Financ. Quant. Anal. **34**(1), 33–55 (1999)
2. Angelov, P.: Evolving Takagi-Sugeno fuzzy systems from data streams (eTS+). In: Angelov, P., Filev, D., Kasabov, N. (eds.) Evolving Intelligent Systems: Methodology and Applications, chap. 2, pp. 21–50. Wiley and IEEE Press, Hoboken, NJ, USA (2010)
3. Angelov, P., Filev, D.: An approach to online identification of Takagi-Sugeno fuzzy models. IEEE Trans. Syst. Man Cybern.—Part B **34**(1), 484–498 (2004)
4. Angelov, P., Filev, D.: Simpl_eTS: a simplified method for learning evolving Takagi-Sugeno fuzzy models. In: IEEE International Conference on Fuzzy Systems, Reno, Nevada, USA, pp. 1068–1073 (2005)
5. Angelov, P., Zhou, X.: Evolving fuzzy systems from data streams in real-time. In: International Symposium on Evolving Fuzzy Systems, Ambleside, Lake District, United Kingdom, pp. 29–35(2006)
6. Azzem, M.F., Hanmandlu, M., Ahmad, N.: Structure identification of generalized adaptive neuro-fuzzy inference systems. IEEE Trans. Fuzzy Syst. **11**(5), 668–681 (2003)
7. Ballini, R., Mendonça, A.R.R., Gomide, F.: Evolving fuzzy modeling in risk analysis. Intell. Syst. Acc. Finan. Manage. **16**, 71–86 (2009)
8. Bera, A., Jarque, C.: Efficient tests for normality, homocedasticity and serial independence of regression residuals: Monte Carlo evidence. Econ. Lett. **7**, 313–318 (1981)

9. Bounhas, M., Hamed, M.G., Prade, H., Serrurier, M., Mellouli, K.: Naive possibilistic classifiers for imprecise or uncertain numerical data. Fuzzy Sets Syst. **239**, 137–156 (2014)
10. Casarin, R., Chang, C., Jimenez-Martin, J., Pérez-Amaral, M.M.T.: Risk management of risk under the basel accord: a Bayesian approach to forecasting value-at-risk of VIX futures. Math. Comput. Simul. **94**, 183–204 (2013)
11. Dovzan, D., Loga, V., Skrjanc, I.: Solving the sales prediction with fuzzy evolving models. In: WCCI 2012 IEEE World Congress on Computational Intelligence, June, Brisbane, Australia, pp. 10–15 (2012)
12. Dovžan, D., Škrjanc, I.: Recursive clustering based on a Gustafson-Kessel algorithm. Evol. Syst. **2**(1), 15–24 (2011)
13. Dovžan, D., Škrjanc, I.: Recursive fuzzy c-means clustering for recursive fuzzy identification of time-varying processes. ISA Trans. **50**(2), 159–169 (2011)
14. Dowd, K., Blake, D.: After var: the theory, estimation and insurance applications of quantile-based risk measures. J. Risk Insur. **73**(2), 193–229 (2006)
15. Dunis, C., Laws, J., Sermpinis, G.: Modeling commodity value-at-risk with high order neural networks. Appl. Finan. Econ. **20**(7), 585–600 (2010)
16. Engle, R.F.: Autoregressive conditional heteroscedasticity with estimates of the variance of United Kingdom inflation. Econometrica **50**(4), 987–1007 (1982)
17. Engle, R.F., Manganelli, S.: Caviar: conditional autoregressive value-at-risk by regression quantiles. J. Bus. Econ. Stat. **22**(4), 367–381 (2004)
18. Ferraro, M.B., Giordani, P.: On possibilistic clustering with repulsion constraints for imprecise data. Inf. Sci. **245**, 63–75 (2013)
19. Filev, D., Georgieva, O.: An extended version of the Gustafson-Kessel algorithm for evolving data streams clustering. In: Angelov, P., Filev, D., Kasabov, N. (eds.) Evolving Intelligent Systems: Methodology and Applications, chap. 12, pp. 273–299. Wiley and IEEE Press, Hoboken, NJ, USA (2010)
20. Haas, M., Mittinik, S., Paolella, M.S.: A new approach to Markov switching GARCH models. J. Finan. Econom. **2**(4), 493–530 (2004)
21. Hartz, C., Mittinik, S., Paolella, M.S.: Accurate value-at-risk forecasting based on the normal-GARCH model. Comput. Stat. Data Anal. **51**(4), 2295–2312 (2006)
22. Kasabov, N.K., Song, Q.: DENFIS: dynamic evolving neural-fuzzy inference system and its application for time-series prediction. IEEE Trans. Fuzzy Syst. **10**(2), 144–154 (2002)
23. Keller, J.M., Gray, M.R., Givens, J.A.: A fuzzy k-nearest neighbor algorithm. IEEE Trans. Syst. Man Cybern. **15**(4), 580–585 (1985)
24. Kohonen, T.: Self-organization and Associative Memory, 3rd edn. Springer, Berlin (1989)
25. Krishnapuram, R., Keller, J.: A possibilistic approach to clustering. IEEE Trans. Fuzzy Syst. **2**(1), 98–110 (1993)
26. Kuester, K., Mittinik, S., Paolella, M.S.: Value-at-risk prediction: a comparison of alternative strategies. J. Finan. Econom. **4**(1), 53–89 (2005)
27. Lemos, A.P., Caminhas, W., Gomide, F.: Fuzzy evolving linear regression trees. Evol. Syst. **2**, 1–14 (2011)
28. Leng, G., McGINNITY, T.M., Prasad, G.: An approach for on-line extraction of fuzzy rules using a self-organizing fuzzy neural network. Fuzzy Sets Syst. **150**(2), 211–243 (2005)
29. Liang, N., Huang, G., Saratchandran, P., Sundararajan, N.: A fast and accurate online sequential learning algorithm for feedforward networks. IEEE Trans. Neural Netw. **17**(6), 1411–1423 (2006)
30. Lima, E., Hell, H., Ballini, R., Gomide, F.: Evolving fuzzy modeling using participatory learning. In: Angelov, P., Filev, D., Kasabov, N. (eds.) Evolving intelligent systems: methodology and applications, chap. 4, pp. 67–86. Wiley and IEEE Press, Hoboken, NJ, USA (2010)
31. Ljung, L.: System Identification: Theory for the User. Prentice-Hall, Englewood Cliffs (1988)
32. Lopez, J.A.: Regulatory evaluation of value-at-risk models. J. Risk **1**, 37–64 (1999)
33. Lughofer, E.D.: FLEXFIS: a robust incremental learning approach for evolving takagi-sugeno fuzzy models. IEEE Trans. Fuzzy Syst. **16**(6), 1393–1410 (2008)

34. Lughofer, E.: Evolving fuzzy systems: methodologies, advances concepts and applications. Springer, Berlin (2011)
35. Luna, I., Ballini, R.: Online estimation of stochastic volatility for asset returns. In: IEEE Computational Intelligence for Financial Engineering and Economics (CIFEr 2012) (2012)
36. Maciel, L., Gomide, F., Ballini, R.: Enhanced evolving participatory learning fuzzy modeling: an application for asset returns volatility forecasting. Evol. Syst. 5(1), 75–88 (2014)
37. Maciel, L., Gomide, F., Ballini, R.: Recursive possibilistic fuzzy modeling. Evolving and autonomous learning systems (EALS). In: IEEE Symposium Series on Computation Intelligence (SSCI), Orlando, FL, pp. 9–16 (2014)
38. Maciel, L., Gomide, F., Ballini, R.: Evolving possibilistic fuzzy modeling for realized volatility forecasting with jumps. IEEE Trans. Fuzzy Syst. (submitted) (2015)
39. Maciel, L., Gomide, F., Ballini, R., Yager, R.: Simplified evolving rule-based fuzzy modeling of realized volatility forecasting with jumps. In: IEEE Workshop on Computational Intelligence for Financial Engineering and Economics (CIFEr 2013), Cingapura, pp. 76–83 (2013)
40. McNeil, A.J., Frey, R.: Estimation of tail-related risk measures for heteroscedastic financial time series: an extreme value approach. J. Empirical Finan. 7(3), 271–300 (2000)
41. Moussa, A.M., Kamdem, J.S., Terraza, M.: Fuzzy value-at-risk and expected shortfall for portfolios with heavy-tailed returns. Econ. Model. 39, 247–256 (2014)
42. Pal, N.R., Pal, K., Keller, J.M., Bezdek, J.C.: A possibilistic fuzzy c-means clustering algorithm. IEEE Trans. Fuzzy Syst. 13(4), 517–530 (2005)
43. Qiao, J., Wang, H.: A self-organising fuzzy neural network and its application to function approximation and forest modeling. Neurocomputing 71(4–6), 564–569 (2008)
44. Riskmetrics, J.P.: Morgan technical documentation, 4th edn. Technical report. J.P. Morgan, New York (1996)
45. Rong, H.J., Sundarajan, N., Huang, G., Zhao, G.: Extended sequential adaptive fuzzy inference system for classification problems. Evol. Syst. 2, 71–82 (2011)
46. Rosa, R., Maciel, L., Gomide, F., Ballini, R.: Evolving hybrid neural fuzzy network for realized volatility forecasting with jumps. In: IEEE Workshop on Computational Intelligence for Financial Engineering and Economics (CIFEr 2014), Londres, vol. 1, pp. 1–8 (2014)
47. Schwarz, G.: Estimating the dimension of model. Ann. Stat. 6(2), 461–464 (1978)
48. Timm, H., Borgelt, C., Doring, C., Kruse, R.: An extension to possibilistic fuzzy cluster analysis. Fuzzy Sets Syst. 147(1), 3–16 (2004)
49. Tung, W.L., Quek, C.: Financial volatility trading using a self-organising neural-fuzzy semantic network and option straddle-based approach. Expert Syst. Appl. 38(5), 4668–4688 (2011)
50. Yager, R.: A model of participatory learning. IEEE Trans. Syst. Man Cybern. 20(5), 1229–1234 (1990)
51. Yager, R.R., Filev, D.P.: Modeling participatory learning as a control mechanism. Int. J. Intell. Syst. 8(3), 431–450 (1993)
52. Yoshida, Y.: An estimation model of value-at-risk portfolio under uncertainty. Fuzzy Sets Syst. 160(22), 3250–3262 (2009)
53. Zmeskal, Z.: Value-at-risk methodology of international index portfolio under soft conditions (fuzzy-stochastic approach). Int. Rev. Finan. Anal. 14(2), 263–275 (2005)

Using Similarity and Dissimilarity Measures of Binary Patterns for the Comparison of Voting Procedures

Janusz Kacprzyk, Hannu Nurmi and Sławomir Zadrożny

Abstract An interesting and important problem of how similar and/or dissimilar voting procedures (social choice functions) are dealt with. We extend our previous qualitative type analysis based on rough sets theory which make it possible to partition the set of voting procedures considered into some subsets within which the voting procedures are indistinguishable, i.e. (very) similar. Then, we propose an extension of those analyses towards a quantitative evaluation via the use of degrees of similarity and dissimilarity, not necessarily metrics and dual (in the sense of summing up to 1). We consider the amendment, Copeland, Dodgson, max-min, plurality, Borda, approval, runoff, and Nanson, voting procedures, and the Condorcet winner, Condorcet loser, majority winner, monotonicity, weak Pareto winner, consistency, and heritage criteria. The satisfaction or dissatisfaction of the particular criteria by the particular voting procedures are represented as binary vectors. We use the Jaccard–Needham, Dice, Correlation, Yule, Russell–Rao, Sockal–Michener, Rodgers–Tanimoto, and Kulczyński measures of similarity and dissimilarity. This makes it possible to gain much insight into the similarity/dissimilarity of voting procedures.

To Ron, Professor Ronald R. Yager, whose highly original and ground breaking ideas, and vision, have shaped and changed research interests of so many of us for years.

J. Kacprzyk (✉) · S. Zadrożny
Systems Research Institute, Polish Academy of Sciences, Ul. Newelska 6,
01-447 Warsaw, Poland
e-mail: kacprzyk@ibspan.waw.pl

J. Kacprzyk · S. Zadrożny
Warsaw School of Information Technology (WIT), Ul. Newelska 6,
01-447 Warsaw, Poland

H. Nurmi
Department of Political Science, University of Turku, 20014 Turku, Finland
e-mail: hnurmi@utu.fi

© Springer International Publishing Switzerland 2017
J. Kacprzyk et al. (eds.), *Granular, Soft and Fuzzy Approaches
for Intelligent Systems*, Studies in Fuzziness and Soft Computing 344,
DOI 10.1007/978-3-319-40314-4_8

141

1 Introduction

In this paper we deal with *voting procedures*, maybe the most intuitively appealing examples of *social choice function*, which are meant to determine the winner of some election in the function of individual votes—cf. for a comprehensive exposure in particular Pitt et al. [1] but also Pitt et al. [2, 3], Arrow, Sen and Suzumura [4], Kelly [5], Plott [6], Schwartz [7], etc.

Basically, we consider the following problem: we have n, $n \geq 2$ individuals who present their testimonies over the set of m, $m \geq 2$, options. The testimonies can be exemplified by *individual preference relations* which are often, also here, binary relations over the set of options, orderings over the set of options. We look for *social choice functions*, or—to be more specific—a voting procedure that would select a set of options that would best reflect the opinions of the whole group, as a function of individual preference relations.

A traditional line of research here has been whether and to which extent the particular voting procedures do or do not satisfy some plausible and reasonable axioms and conditions, maybe best exemplified by the famous Arrows theorem, and so many paradoxes of voting. We will not deal with this, for details cf. Arrow [8], Gibbard [9], Kelly [10], May [11], Nurmi [12], Riker [13], Satterthwaite [14], etc.

We will deal with an equally important, or probably practically more important, problem of how similar or dissimilar the particular voting procedures are. This was discussed in Nurmi's [12] book, cf. also Baigent [15], Elkind, Faliszewski and Slinko [16], McCabe-Dansted and Slinko [17], Richelson [18], etc.

In this paper we will deal with the above mentioned problem of how to measure the similarity and dissimilarity of voting procedures. First, we will take into account only a subset of well known voting procedures. Then, we will employ the idea of a qualitative similarity (and its related dissimilarity) analysis of voting procedures proposed by Fedrizzi, Kacprzyk and Nurmi [19] in which Pawlak's *rough sets* (cf. Pawlak [20, 21], cf. also Pawlak and Skowron [22]), have been used. Then, we will use the idea of the recent approach proposed by Kacprzyk, Nurmi and Zadrożny [23] in which the above mentioned more qualitative rough sets based analysis has been extended with a quantitative analysis by using the Hamming and Jaccard-Needham similarity indexes.

This paper is a further extension of Kacprzyk, Nurmi and Zadrożny [23]. Basically, we consider some other more popular similarity (and their related dissimilarity) measures:

- Jaccard-Needham (to repeat, for completeness, the results already obtained for this measure in [23]),
- Dice,
- correlation,
- Yule,
- Russell–Rao,
- Sockal–Michener,

- Rogers–Tanimoto, and
- Kulczyński—cf. Tubbs [24] for details.

Notice that these measure are just a small subset of a multitude of similarity measures known in the literature, cf. Choi, Cha and Tappert [25]. Moreover, in this paper we limit our attention to those similarity measures which, first of all, take in values in [0, 1], and the corresponding dissimilarity measures of which are dual in the sense that their values add up to 1, which is not the case for all measures.

Notice that this approach is different both conceptually and technically from the approach by Kacprzyk and Zadrożny [26, 27] in which some distinct classes of voting procedures are determined using the concept of Yager's [28] ordered weighted averaging (OWA) aggregation operator (cf. Yager and Kacprzyk [29], Yager, Kacprzyk and Beliakov [30]), and the change of the order of variables to be aggregated and the type of weights (i.e. the aggregation behavior) determines various classes of voting procedures.

2 Foundations of the Theory of Rough Sets

Rough sets were proposed in the early 1980s by Pawlak [20], and then extensively developed by Pawlak [21], Polkowski (e.g., [31]), Skowron (e.g., [22, 32, 33]), Słowiński (e.g., [34]), etc. and their collaborators. It is a conceptually simple and intuitively appealing tool for the representation and processing of imprecise knowledge when the classes into which the objects are to be classified are imprecise but can be approximated by precise sets, from the above and below.

Here we will just briefly recall some basic concepts and properties of rough sets theory which may be useful for our purpose, and for more detail, cf. Pawlak [20], [21], Polkowski (e.g., [31]), Skowron (e.g., [22], Pawlak and Skowron [32, 35], Pawlak et al. [33]), and Greco et al. (e.g., [34]) etc. to just list a few.

Let $U = \{u\}$ be a *universe of discourse*. It can usually be partitioned in various ways into a family $\mathbf{R}$ of partitionings, or equivalence relations defined on U. A *knowledge base*, denoted by K, is the pair $K = (U, \mathbf{R})$. Let now $\mathbf{P}$ be a non-empty subset of $\mathbf{R}, \mathbf{P} \subset \mathbf{R}, \mathbf{P} \neq \emptyset$. Then, the intersection of all equivalence relations (or partitionings) in $\mathbf{P}$, which is also an equivalence relation, is called an *indiscernibility relation* over $\mathbf{P}$ and is denoted by $IND(\mathbf{P})$.

The family of its equivalence classes is termed the $\mathbf{P}$-basic knowledge about U in K and it represents all that can be said about the elements of U under $\mathbf{P}$. Therefore, one cannot classify the elements of U any deeper than to the equivalence classes of $IND(\mathbf{P})$. For instance, if for some $U, \mathbf{P} = \{R_1, R_2\}$ such that R_1 partitions the objects into the classes labeled "heavy" and "lightweight", and R_2 partitions into the classes labeled "black" and "white", then all that can be said about any element of U is that it belongs to one of: "heavy-and-black", "heavy-and-white", "lightweight-and-black", "lightweight-and-white".

Equivalence classes of $IND(\mathbf{P})$ are called the *basic categories (concepts) of knowledge P*. If $Q \in \mathbf{R}$, that is, Q is an equivalence relation on U, then its equivalence classes are called the *Q-elementary categories (concepts) of knowledge* $\mathbf{R}$.

If $X \subset U$, and R is an equivalence relation on U, then X is called *R-definable* or *R-exact* if it is a union of some R-elementary categories (R-basic categories); otherwise, it is called R-rough.

Rough sets can be approximately defined by associating with any $X \subset U$ and any equivalence relation R on U the following two sets (U/R denotes the set of all equivalence relations of R):

- a *lower approximation* of X:

$$R_L X = \bigcup \{Y \in U/R \mid Y \subset X\} \tag{1}$$

- an *upper approximation* of X:

$$R_U X = \bigcup \{Y \in U/R \mid Y \cap X \neq \emptyset\} \tag{2}$$

and a *rough set* is defined as the pair (R_L, R_U).

The lower approximation yields the classes of R which are subsets of X, i,e, contains those elements which are necessarily also elements of X, while the upper approximation yields those classes of R which have at least one common element with X.

For our purposes two concepts related to the reduction of knowledge are crucial. First, for a family of equivalence relations $\mathbf{R}$ on U, one of its elements, Z, is called *dispensable* in $\mathbf{R}$ if

$$IND(\mathbf{R}) = IND(\mathbf{R} \setminus \{Z\}) \tag{3}$$

and otherwise it is called *indispensable*. If each Z in $\mathbf{R}$ is indispensable, then $\mathbf{R}$ is called *independent*.

For a family of equivalence relations, $\mathbf{R}$, and its subfamily, $\mathbf{Q} \subset \mathbf{R}$, if:

- $\mathbf{Q}$ is independent, and
- $IND(\mathbf{Q}) = IND(\mathbf{R})$,

then $\mathbf{Q}$ is called a *reduct* of $\mathbf{R}$; clearly, it need not be unique.

The *core* of $\mathbf{R}$ is the set of all indispensable equivalence relations in $\mathbf{R}$, and is the intersection of all reducts of $\mathbf{R}$—cf. Pawlak [20].

From the point of view of knowledge reduction, the core consists of those classifications (equivalence relations) which are the most essential in the knowledge available in that no equivalence relation that belongs to the core can be discarded in the knowledge reduction process without distorting the knowledge itself. A reduct yields a set of equivalence relations which is sufficient for the characterization of knowledge available without losing anything relevant.

In this paper our analysis is in terms of indiscernibility relations; for the concept of a discernibility relation, cf. Yao and Zhao [36].

3 A Comparison of Voting Procedures Using Rough Sets

The problem of comparison and evaluation of voting procedures (social choice functions) is very important and has been widely studied in the literature, cf. Richelson [18], Straffin [37], Nurmi [12], to name a few.

A simple, intuitively appealing, rough set based approach, was proposed by Fedrizzi, Kacprzyk and Nurmi [19]. It was more qualitative, and was extended to include more quantitative aspects by Kacprzyk, Nurmi and Zadrożny [23]. We will now briefly recall this approach since it will provide a point of departure for this paper.

We assume that we have 13 popular voting procedures:

1. Amendment: proposals (options) are paired (compared) with the status quo. If a variation on the proposal is introduced, then it is paired with this proposal and voted on as an amendment prior to the final vote. Then, if the amendment succeeds, the amended proposal is eventually paired with the status quo in the final vote, otherwise, the amendment is eliminated prior to the final vote.
2. Copeland: selects the option with the largest so-called Copeland score which is the number of times an option beats other options minus the number of times this option loses to other options, both in pairwise comparisons.
3. Dodgson: each voter gives a rank ordered list of all options, from the best to the worst, and the winner is the option for which the minimum number of pairwise exchanges (added over all candidates) is needed before they all become a Condorcet winner, i.e. defeat all other options in pairwise comparisons with a majority of votes.
4. Schwartz: selects the set of options over which the collective majority preferences are cyclic and the entire cycle is preferred over the other options; it is the single element in case there is a Condorcet winner, otherwise it consists of several options.
5. Max-min: selects the option for which the minimal support in all pairwise comparisons is the largest.
6. Plurality: each voter selects one option (or none in the case of abstention), and the options with the most votes win.
7. Borda: each voter provides a linear ordering of the options which are assigned a score (the so-called Borda score) as follows: if there are n candidates, $n - 1$ points are given to the first ranked option, $n - 2$ to the second ranked, etc., and these numbers are summed up for each option to end up with the Borda score for this option, and the option(s) with the highest Borda score win(s).
8. Approval: each voter selects a subset of the candidate options and the option(s) with the most votes is/are the winner(s).
9. Black: selects either the Condorcet winner, i.e. an option that beats or ties with all others in pairwise comparisons, when it exists, and the Borda count winner (as described above) otherwise.
10. Runoff: the option ranked first by more than a half of the voters is chosen if one exists. Otherwise, the two options ranked first by more voters than any

other option are compared with each other and the winner is the one ranked first (among the remaining options) by more voters than the other option.

11. Nanson: we iteratively use the Borda count, at each step dropping the candidate with the smallest score (majority); in fact, this is sometimes called a modified version of the Nanson rule, cf. Fishburn [38],

12. Hare: the ballots are linear orders over the set of options, and we repeatedly delete the options which receive the lowest number of first places in the votes, and the option(s) that remain(s) are declared as the winner(s).

13. Coombs: each voter rank orders all of the options, and if one option is ranked first by an absolute majority of the voters, then it is the winner. Otherwise, the option which is ranked last by the largest number of voters is eliminated, and the procedure is repeated.

What concerns the criteria against which the above mentioned voting procedures are compared, we use some basic and popular ones presented in the classic Nurmi's [12] book. More specifically, we will consider 7 criteria the voting procedures are to satisfy:

1. A—Condorcet winner,
2. B—Condorcet loser,
3. C—majority winner,
4. D—monotonicity,
5. E—weak Pareto winner,
6. F—consistency, and
7. G—heritage,

the essence of which can be summarized as:

1. Condorcet winner: if an option beats each other option in pairwise comparisons, it should always win.
2. Condorcet loser: if an option loses to each other option in pairwise comparisons, it should always loose.
3. Majority winner: if there exists a majority (at least a half) that ranks a single option as the first, higher than all other candidates, that option should win.
4. Monotonicity: it is impossible to cause a winning option to lose by ranking it higher, or to cause a losing option to win by ranking it lower.
5. Weak Pareto winner: whenever all voters rank an option higher than another option, the latter option should never be chosen.
6. Consistency criterion: if the electorate is divided in two and an option wins in both parts, it should win in general.
7. Heritage: if an option is chosen from the entire set of options using a particular voting procedure, then it should also be chosen from all subsets of the set of options (to which it belongs) using the same voting procedure and under the same preferences.

We start with a illustrative account of which voting procedure satisfies which criteria("0" stands for "does not satisfy", and "1" stands for "satisfies") which is

Table 1 Satisfaction of 7 criteria by 13 voting procedures

Voting procedure	Criteria						
	A	B	C	D	E	F	G
Amendment	1	1	1	1	0	0	0
Copeland	1	1	1	1	1	0	0
Dodgson	1	0	1	0	1	0	0
Schwartz	1	1	1	1	0	0	0
Max-min	1	0	1	1	1	0	0
Plurality	0	0	1	1	1	1	0
Borda	0	1	0	1	1	1	0
Approval	0	0	0	1	0	1	1
Black	1	1	1	1	1	0	0
Runoff	0	1	1	0	1	0	0
Nanson	1	1	1	0	1	0	0
Hare	0	1	1	0	1	0	0
Coombs	0	1	1	0	1	0	0

presented in Table 1; the 13 voting procedures correspond to the rows while the 7 criteria correspond to the columns, here and in next tables.

Though the data shown in Table 1 can be immediately used for the comparison of the 17 voting procedures against the 7 criteria by a simple pairwise comparison of rows, a natural attempt is to find first if, under the information available in that table, all the 17 voting procedures are really different, and if all the 7 criteria as really needed for a meaningful comparison.

Quite a natural, simple and intuitively appealing approach was proposed in this respect by Fedrizzi et al. [19] using rough sets. We will present below its essence.

4 Simplification of Information on the Voting Procedures and Criteria to Be Fulfilled

We will now show the essence of Fedrizzi et al. [19] approach based on the application of some elements of rough sets theory, briefly presented in Sect. 2, to simplify information in the source Table 1. We will basically consider crucial properties or attributes of the voting procedures that will make it possible to merge them into one (class of) voting procedure under a natural condition that they satisfy the same properties, i.e. the criteria assumed.

First, one can see that the amendment procedure and Schwartz' choice function have identical properties in Table 1, so one can be deleted and similarly for Copeland's and Black's choice functions, the runoff, Hare's and Coombs' choice functions. We obtain therefore Table 2.

Table 2 Satisfaction of 7 criteria by 9 equivalent (classes of) voting procedures

Voting procedure	Criteria						
	A	B	C	D	E	F	G
Amendment	1	1	1	1	0	0	0
Copeland	1	1	1	1	1	0	0
Dodgson	1	0	1	0	1	0	0
Max-min	1	0	1	1	1	0	0
Plurality	0	0	1	1	1	1	0
Borda	0	1	0	1	1	1	0
Approval	0	0	0	1	0	1	1
Runoff	0	1	1	0	1	0	0
Nanson	1	1	1	0	1	0	0

So that we have 9 "really different" (classes of) voting procedures:

1. Amendment (which stands now for Amendment and Schwartz),
2. Copeland (which stands now for Copeland and Black),
3. Dodgson,
4. Max-min,
5. Plurality,
6. Borda,
7. Approval,
8. Runoff (which stands now for Runoff, and Hare and Coombs).
9. Nanson.

Now, we look for the *indispensable criteria*, cf. Sect. 2 which boils down to that if we take into account that each attribute (which corresponds to a criterion) generates such an equivalence relation, then to the same class there belong those voting procedures that fulfill those criteria, and to another class those which do not. This can be done by eliminating the criteria one by one and finding out whether the voting procedures can be discerned from each other in terms of the remaining criteria.

Therefore, if we start from Table 2, by eliminating criterion A we get Table 3.

The two last rows of Table 3 are identical and to distinguish those two last rows, i.e. Runoff and Nanson, criterion A is necessary, i.e. criterion A is indispensable.

And, analogously, we delete criterion B and obtain Table 4.

The Copeland and Max-Min procedures become indistinguishable so that criterion B is indispensable.

Next, the elimination of criterion C leads to Table 5.

All rows in Table 5 are different so that criterion C is unnecessary to differentiate between those voting functions, and we can conclude that C is dispensable.

Table 3 Elimination of criterion A from Table 2

Voting procedure	Criteria					
	B	C	D	E	F	G
Amendment	1	1	1	0	0	0
Copeland	1	1	1	1	0	0
Dodgson	0	1	0	1	0	0
Max-min	0	1	1	1	0	0
Plurality	0	1	1	1	1	0
Borda	1	0	1	1	1	0
Approval	0	0	1	0	1	1
Runoff	1	1	0	1	0	0
Nanson	1	1	0	1	0	0

Table 4 Elimination of criterion B from Table 2

Voting procedure	Criteria					
	A	C	D	E	F	G
Amendment	1	1	1	0	0	0
Copeland	1	1	1	1	0	0
Dodgson	1	1	0	1	0	0
Max-min	1	1	1	1	0	0
Plurality	0	1	1	1	1	0
Borda	0	0	1	1	1	0
Approval	0	0	1	0	1	1
Runoff	0	1	0	1	0	0
Nanson	1	1	0	1	0	0

Further, we delete criterion D and obtain Table 6. The Copeland and Nanson choice functions are now indistinguishable which means that criterion D is indispensable.

Now, we eliminate criterion E and get Table 7.

Two uppermost rows are now identical so that criterion E is needed, i.e. it is indispensable.

Next, criterion F is eliminated as shown in Table 8 in which no pair of rows is identical so that criterion F is dispensable.

Finally, criterion G is eliminated which is shown in Table 9. We can see that all rows are different so that we can conclude that criterion G is dispensable.

It is easy to notice that the core is the set of indispensable criteria, i.e. {A, B, D, E}, and the reduct is in this case both unique and also equal to {A, B, D, E}. That is,

Table 5 Elimination of criterion C from Table 2

Voting procedure	Criteria					
	A	B	D	E	F	G
Amendment	1	1	1	0	0	0
Copeland	1	1	1	1	0	0
Dodgson	1	0	0	1	0	0
Max-min	1	0	1	1	0	0
Plurality	0	0	1	1	1	0
Borda	0	1	1	1	1	0
Approval	0	0	1	0	1	1
Runoff	0	1	0	1	0	0
Nanson	1	1	0	1	0	0

Table 6 Elimination of criterion D from Table 2

Voting procedure	Criteria					
	A	B	C	E	F	G
Amendment	1	1	1	0	0	0
Copeland	1	1	1	1	0	0
Dodgson	1	0	1	1	0	0
Max-min	1	0	1	1	0	0
Plurality	0	0	1	1	1	0
Borda	0	1	0	1	1	0
Approval	0	0	0	0	1	1
Runoff	0	1	1	1	0	0
Nanson	1	1	1	1	0	0

we need just that set of criteria to distinguish the particular voting procedures from each other (naturally, under the set of criteria assumed).

We can then consider the reduct (or core). In Table 10 we show which criteria are indispensable in the sense that if we do not take them into account, the two or more rows (corresponding to the respective voting procedures) become indistinguishable. For example, without criterion E, Amendment and Copeland would be indistinguishable, without D, Copeland and Nanson would be indistinguishable, without B, Copeland and Max-Min would be indistinguishable, etc.

Table 10 expresses the most crucial properties or criteria of the voting procedures in the sense that the information it conveys would be sufficient to restore all information given in the source Table 2. Therefore, for an "economical" characterization of the voting procedures, we can use the values of the criteria given in Table 10 and present the results as in Table 11 where the subscripts of the particular criteria stand

Table 7 Elimination of criterion E from Table 2

Voting procedure	Criteria					
	A	B	C	D	F	G
Amendment	1	1	1	1	0	0
Copeland	1	1	1	1	0	0
Dodgson	1	0	1	0	0	0
Max-min	1	0	1	1	0	0
Plurality	0	0	1	1	1	0
Borda	0	1	0	1	1	0
Approval	0	0	0	1	1	1
Runoff	0	1	1	0	0	0
Nanson	1	1	1	0	0	0

Table 8 Elimination of criterion F from Table 2

Voting procedure	Criteria					
	A	B	C	D	E	G
Amendment	1	1	1	1	0	0
Copeland	1	1	1	1	1	0
Dodgson	1	0	1	0	1	0
Max-min	1	0	1	1	1	0
Plurality	0	0	1	1	1	0
Borda	0	1	0	1	1	0
Approval	0	0	0	1	0	1
Runoff	0	1	1	0	1	0
Nanson	1	1	1	0	1	0

for the values they take on, for instance, to most economically characterize Amendment, the A, B and D should be 1 and E should be 0, etc.

This is, however, not yet the most economical characterization but this issues will not be dealt with here and we refer the interested reader to Fedrizzi, Kacprzyk and Nurmi [19] or Kacprzyk, Nurmi and Zadrożny [23] to find that the minimal (most economical) characterization of the voting procedures in term of information given in Table 2 can be portrayed as shown in Table 12.

This is a very compact representation which is due to the very power of rough sets theory.

Table 9 Elimination of criterion F from Table 2

Voting procedure	Criteria					
	A	B	C	D	E	F
Amendment	1	1	1	1	0	0
Copeland	1	1	1	1	1	0
Dodgson	1	0	1	0	1	0
Max-min	1	0	1	1	1	0
Plurality	0	0	1	1	1	1
Borda	0	1	0	1	1	1
Approval	0	0	0	1	0	1
Runoff	0	1	1	0	1	0
Nanson	1	1	1	0	1	0

Table 10 Satisfaction of the criteria belonging to the core by the particular voting procedures

Voting procedure	Criteria			
	A	B	D	E
Amendment	1	1	1	0
Copeland	1	1	1	1
Dodgson	1	0	0	1
Max-min	1	0	1	1
Plurality	0	0	1	1
Borda	0	1	1	1
Approval	0	0	1	0
Runoff	0	1	0	1
Nanson	1	1	0	1

Table 11 An economical characterization of the voting procedures shown in Table 10

$A_1 B_1 D_1 E_0 \longrightarrow$ Amendment
$A_1 B_1 D_1 E_1 \longrightarrow$ Copeland
$A_1 B_0 D_0 E_1 \longrightarrow$ Dodgson
$A_1 B_0 D_1 E_1 \longrightarrow$ Max-min
$A_0 B_0 D_1 E_1 \longrightarrow$ Plurality
$A_0 B_1 D_1 E_1 \longrightarrow$ Borda
$A_0 B_0 D_1 E_0 \longrightarrow$ Approval
$A_0 B_1 D_0 E_1 \longrightarrow$ Runoff
$A_1 B_1 D_0 E_1 \longrightarrow$ Nanson

Table 12 The minimal (most economical) characterization of the voting procedures shown in Table 10

$A_1E_0 \longrightarrow$ Amendment	
$A_1B_1D_1E_1 \longrightarrow$ Copeland	
$B_0D_0 \longrightarrow$ Dodgson	
$A_1B_0D_1 \longrightarrow$ Max-min	
$A_0B_0E_1 \longrightarrow$ Plurality	
$A_0B_1D_1 \longrightarrow$ Borda	
$A_0E_0 \longrightarrow$ Approval	
$A_0D_0 \longrightarrow$ Runoff	
$A_1B_1D_0 \longrightarrow$ Nanson	

5 Similarity and Dissimilarity of the Voting Procedures: A Quantitative Approach Based on Similarity and Dissimilarity Measures for Binary Patterns

As it could be seen from the previous section, a rough sets based analysis has made it possible to find a smaller subset of all the choice functions considered such that choice functions merged could have been meant as similar. This, rather qualitative result, is clearly the first step. The next steps towards a more quantitative analysis can be made, using elements of rough sets theory, by using some indiscernibility analyses. This was proposed by Fedrizzi, Kacprzyk and Nurmi [19], and then extended by Kacprzyk, Nurmi and Zadrożny [23]. We will not deal with this approach and refer the interested reader to the above mentioned papers.

In this paper we will approach the problem of measuring the similarity and dissimilarity in a more quantitative way, using some similarity and dissimilarity measures, but going beyond the classic Hamming and Jaard-Needham measures used in Kacprzyk, Nurmi and Zadrożny [23].

We take again as the point of departure the characterization of the voting procedures as shown in Table 2, that is, just after the reduction of identical rows in Table 1, but—to better show the generality of our approach—without all further reductions (or a representation size reduction) as proposed later on and presented in Tables 3–10.

The data sets involved are in fact binary patterns and there is a multitude of similarity/dissimilarity measures for binary patterns but we will here concentrate on the measures given by Tubbs [24] which are useful in matching binary patterns in pattern recognition. We will follow to a large extent Tubb's notation.

A binary vector Z of dimension N is defined as:

$$Z = (z_1, z_2, \ldots, z_N) \tag{4}$$

where $z_i \in \{0, 1\}$, $\forall i \in \{1, 2, \ldots, N\}$.

The set of all N-dimensional binary vectors is denoted by Ω, the *unit binary vector*, $I \in \Omega$, is a binary vector such that $z_i = 1, \forall i \in \{1, 2, \ldots, N\}$, and the *complement* of a binary vector $Z \in \Omega$ is $\overline{Z} = I - Z$.

The *magnitude* of a binary vector $Z \in \Omega$ is

$$|Z| = \sum_{i=1}^{N} z_i \tag{5}$$

that is, the number of elements which are equal to 1.

If we have two binary vectors, $X, Y \in \Omega$, then we denote by $S_{i,j}(X, Y)$ the number of matches of i in vector X and j in vector Y, $i, j \in \{0, 1\}$. That is, if we have two vectors:

$$X = [0, 1, 1, 0, 1, 0, 0, 1, 1, 0]$$

$$Y = [1, 1, 0, 0, 1, 1, 0, 0, 1, 0]$$

then we have:

$$S_{00}(X, Y) = 3$$
$$S_{01}(X, Y) = 2$$
$$S_{10}(X, Y) = 2$$
$$S_{11}(X, Y) = 3$$

Formally, we can define those measures as follows. First, for vectors $X = (x_1, x_2, \ldots, x_N)$ and $Y = (y_1, y_2, \ldots, y_N)$:

$$v_{ij} = \begin{cases} 1 & \text{if } x_i = y_j \\ 0 & \text{otherwise} \end{cases} \tag{6}$$

$$v_{ij}^k(X, Y) = \begin{cases} 1 & \text{if } x_k = i \text{ and } y_k = j \\ 0 & \text{otherwise} \end{cases} \tag{7}$$

then

$$S_{ij}(X, Y) = \sum_{k=1}^{N} (v_{00}^k(X, Y) + v_{01}^k(X, Y) + v_{10}^k(X, Y) + v_{11}^k(X, Y)) \tag{8}$$

One can easily notice that

$$S_{00}(X, Y) = \overline{X} \times \overline{Y}^T \tag{9}$$
$$S_{11}(X, Y) = X \times Y^T \tag{10}$$

where "$\times$" denotes the product of the matrices.

Following the notation of Tubbs [24], the S_{ij}'s, $i,j \in \{0,1\}$, can be used to define many well known measures of similarity and dissimilarity, and we will consider here the following ones (we follow here the source terminology from that paper but in the literature sometimes slightly different names are used):

- Jaccard–Needham,
- Dice,
- correlation,
- Yule,
- Russell–Rao,
- Sockal–Michener,
- Rodgers–Tanimoto, and
- Kulczyński.

These measures, both of *similarity* $S(X, Y)$, and their corresponding measures of *dissimilarity*, $D(X, Y)$, are defined in terms of $S_{ij}(X, Y)$ as follows (we omit the arguments (X, Y), for brevity):

- Jaccard–Needham:

$$S_{J-N} = \frac{S_{11}}{S_{11} + S_{10} + S_{01}} \tag{11}$$

$$D_{J-N} = \frac{S_{10} + S_{01}}{S_{11} + S_{10} + S_{01}} \tag{12}$$

- Dice

$$S_D = \frac{2S_{11}}{2S_{11} + S_{10} + S_{01}} \tag{13}$$

$$D_D = \frac{S_{10} + S_{01}}{2S_{11} + S_{10} + S_{01}} \tag{14}$$

- Correlation

$$S_C = \frac{1}{\sigma}(S_{11}S_{00} - S_{10}S_{01}) \tag{15}$$

$$D_C = \frac{1}{2} - \frac{1}{2\sigma}(S_{11}S_{00} - S_{10}S_{01}) \tag{16}$$

where

$$\sigma = \sqrt{(S_{10} + S_{11})(S_{01} + S_{00})(S_{11} + S_{01})(S_{00} + S_{10})}; \tag{17}$$

- Yule

$$S_Y = \frac{S_{11}S_{00} - S_{10}S_{01}}{S_{11}S_{00} + S_{10}S_{01}} \tag{18}$$

$$D_Y = \frac{S_{10}S_{01}}{S_{11}S_{00} + S_{10}S_{01}} \tag{19}$$

- Russell–Rao

$$S_{R-R} = \frac{S_{11}}{N} \tag{20}$$

$$D_{R-R} = \frac{N - S_{11}}{N} \tag{21}$$

- Sokal–Michener

$$S_{S-M} = \frac{S_{11} + S_{00}}{N} \tag{22}$$

$$D_{S-M} = \frac{S_{10} + S_{01}}{N} \tag{23}$$

- Rogers–Tanimoto

$$S_{R-T} = \frac{S_{11} + S_{00}}{S_{11} + S_{00} + 2S_{10} + 2S_{01}} \tag{24}$$

$$D_{R-T} = \frac{2S_{10} + 2S_{01}}{S_{11} + S_{00} + 2S_{10} + 2S_{01}} \tag{25}$$

- Kulczyński

$$S_K = \frac{S_{11}}{S_{10} + S_{01}} \tag{26}$$

$$D_K = \frac{S_{10} + S_{01} - S_{11} + N}{S_{10} + S_{01} + N} \tag{27}$$

Notice that though not all similarity measures employed are normalized, their respective dissimilarity measures are all normalized to the unit interval [0, 1] which is usually welcome in applications, also in our context. On the other hand, not all the measures exhibit the metric property but this will not be discussed in this paper as the importance of this property is not clear from a practical point of view.

Now, we will use these measures to the evaluation of similarity and dissimilarity of the voting procedures employed in our paper.

We will use as the point of departure the binary matrix given in Table 2 which shows the satisfaction (= 1) or a lack of satisfaction (= 0) of the A, B, C, D, E, F, G

Table 13 Satisfaction of 7 criteria by 9 equivalent (classes of) voting procedures, cf. Table 2

Voting procedure	Criteria						
	A	B	C	D	E	F	G
Amendment	1	1	1	1	0	0	0
Copeland	1	1	1	1	1	0	0
Dodgson	1	0	1	0	1	0	0
Max-min	1	0	1	1	1	0	0
Plurality	0	0	1	1	1	1	0
Borda	0	1	0	1	1	1	0
Approval	0	0	0	1	0	1	1
Runoff	0	1	1	0	1	0	0
Nanson	1	1	1	0	1	0	0

criteria by the 9 (classes of) voting procedures, and this table will be repeated for convenience in Table 13.

Now, we will calculate S_{ij}, $i, j \in \{0, 1\}$, according to (6)–(8), for the particular pairs of 9 voting procedures which will be presented in Table 14 the entries of which are given as $[S_{00}, S_{01}, S_{10}, S_{11}]$, for each pair.

Following (11)–(27) and taking as the point of departure the values of $[S_{00}, S_{01}, S_{10}, S_{11}]$ shown in Table 14, we can calculate the values of the particular similarity and dissimilarity indexes calculated using the methods of:

- Jaccard-Needham,
- Dice,
- Correlation,
- Yule,
- Russell–Rao,
- Sockal–Michener,
- Rodgers–Tanimoto, and
- Kulczyński,

which are shown in the consecutive Tables 15, 16, 17, 18, 19, 20, 21 and 22.

The results concerning the similarity and dissimilarity of the voting procedures with respect to 7 widely accepted criteria that have been obtained by using a set of popular and highly recommended similarity and dissimilarity measures for binary patterns, presented in Tables 15–22, provide a lot of insight that can be very much useful for both social choice and voting theorists. They can also be of relevance for people involved in a more practical task of choosing or even developing a proper voting system in a particular situation. Such an analysis would have been too specific for the purpose of this paper in which a new method is proposed.

To briefly summarize the results obtained, we can say that the quantitative analysis of similarity and dissimilarity via the measures employed in this section, i.e.

Table 14 Values of $[S_{00}, S_{01}, S_{10}, S_{11}]$ for the particular pairs of the voting procedures

Voting procedure	Voting procedure								
	Amendment	Copeland	Dodgson	Max-min	Plurality	Borda	Approval	Runoff	Nanson
Amendment	–	[2, 1, 0, 4]	[2, 1, 2, 2]	[2, 1, 1, 3]	[1, 2, 2, 2]	[1, 2, 2, 2]	[1, 2, 3, 1]	[2, 1, 2, 2]	[2, 1, 1, 3]
Copeland		–	[2, 0, 2, 3]	[2, 0, 1, 4]	1, 1, 2, 3	[1, 1, 2, 3]	[0, 2, 4, 1]	[2, 0, 2, 3]	[2, 0, 1, 4]
Dodgson			–	[3, 1, 0, 3]	[2, 2, 1, 2]	[1, 3, 2, 1]	[1, 3, 2, 1]	[2, 2, 2, 1]	[3, 1, 0, 3]
Max-min				–	[2, 1, 1, 3]	[1, 2, 2, 2]	[1, 2, 3, 1]	[2, 1, 2, 2]	[2, 1, 1, 3]
Plurality					–	[2, 1, 1, 3]	[2, 1, 2, 2]	[2, 1, 2, 2]	[1, 2, 2, 2]
Borda						–	[2, 1, 2, 2]	[2, 1, 2, 2]	[1, 2, 2, 2]
Approval							–	[1, 3, 3, 0]	[0, 4, 3, 0]
Runoff								–	[3, 1, 0, 3]
Nanson									–

Table 15 Values of [degree of similarity, degree of dissimilarity] for the particular pairs of voting procedures using the Jaccard–Needham measures of similarity (11) and dissimilarity (12)

Voting procedure	Voting procedure								
	Amendment	Copeland	Dodgson	Max-min	Plurality	Borda	Approval	Runoff	Nanson
Amendment	–	[0.80, 0.20]	[0.40, 0.60]	[0.60, 0.40]	[0.33, 0.67]	[0.33, 0.67]	[0.17, 0.83]	[0.40, 0.60]	[0.60, 0.40]
Copeland		–	[0.60, 0.40]	[0.80, 0.20]	[0.50, 0.50]	[0.50, 0.50]	[0.14, 0.86]	[0.60, 0.40]	[0.80, 0.20]
Dodgson			–	[0.75, 0.25]	[0.40, 0.60]	[0.17, 0.83]	[0.00, 1.00]	[0.50, 0.50]	[0.75, 0.25]
Max-min				–	[0.60, 0.40]	[0.33, 0.67]	[0.17, 0.83]	[0.40, 0.60]	[0.60, 0.40]
Plurality					–	[0.60, 0.40]	[0.40, 0.60]	[0.40, 0.60]	[0.33, 0.67]
Borda						–	[0.40, 0.60]	[0.40, 0.60]	[0.33, 0.67]
Approval							–	[0.00, 1.00]	[0.00, 1.00]
Runoff								–	[0.75, 0.25]
Nanson									–

Table 16 Values of [degree of similarity, degree of dissimilarity] for the particular pairs of voting procedures using the Dice measures of similarity (13) and dissimilarity (14)

Voting procedure	Voting procedure								
	Amendment	Copeland	Dodgson	Max-min	Plurality	Borda	Approval	Runoff	Nanson
Amendment	-	[0.89, 0.11]	[0.57, 0.43]	[0.75, 0.25]	[0.50, 0.50]	[0.50, 0.50]	[0.29, 0.71]	[0.57, 0.43]	[0.75, 0.25]
Copeland		-	[0.75, 0.25]	[0.89, 0.11]	[0.67, 0.33]	[0.67, 0.33]	[0.25, 0.75]	[0.75, 0.25]	[0.89, 0.11]
Dodgson			-	[0.86, 0.14]	[0.57, 0.43]	[0.29, 0.71]	[0.00, 1.00]	[0.67, 0.33]	[0.86, 0.14]
Max-min				-	[0.75, 0.25]	[0.50, 0.50]	[0.29, 0.71]	[0.57, 0.43]	[0.75, 0.25]
Plurality					-	[0.75, 0.25]	[0.57, 0.43]	[0.57, 0.43]	[0.50, 0.50]
Borda						-	[0.57, 0.43]	[0.57, 0.43]	[0.50, 0.50]
Approval							-	[0.00, 1.00]	[0.00, 1.00]
Runoff								-	[0.86, 0.14]
Nanson									-

Table 17 Values of [degree of similarity, degree of dissimilarity] for the particular pairs of voting procedures using the correlation measures of similarity (15) and dissimilarity (16)

Voting procedure	Voting procedure								
	Amendment	Copeland	Dodgson	Max-min	Plurality	Borda	Approval	Runoff	Nanson
Amendment	–	[0.73, 0.13]	[0.17, 0.42]	[0.42, 0.29]	[-0.17, 0.58]	[-0.17, 0.58]	[-0.42, 0.71]	[0.17, 0.42]	[0.42, 0.29]
Copeland		–	[0.55, 0.23]	[0.73, 0.13]	[0.09, 0.45]	[0.09, 0.45]	[-0.73, 0.87]	[0.55, 0.23]	[0.73, 0.13]
Dodgson			–	[0.75, 0.12]	[0.17, 0.42]	[-0.42, 0.71]	[-0.75, 0.88]	[0.42, 0.29]	[0.75, 0.12]
Max-min				–	[0.42, 0.29]	[-0.17, 0.58]	[-0.42, 0.71]	[0.17, 0.42]	[0.42, 0.29]
Plurality					–	[0.42, 0.29]	[0.17, 0.42]	[0.17, 0.42]	[-0.17, 0.58]
Borda						–	[0.17, 0.42]	[0.17, 0.42]	[-0.17, 0.58]
Approval							–	[-0.75, 0.88]	[-1.00, 1.00]
Runoff								–	[0.75, 0.12]
Nanson									–

Table 18 Values of [degree of similarity, degree of dissimilarity] for the particular pairs of voting procedures using the Yule measures of similarity (18) and dissimilarity (19)

Voting procedure	Voting procedure								
	Amendment	Copeland	Dodgson	Max-min	Plurality	Borda	Approval	Runoff	Nanson
Amendment	–	[1.00, 0.00]	[0.33, 0.33]	[0.71, 0.14]	[-0.33, 0.67]	[-0.33, 0.67]	[-0.71, 0.86]	[0.33, 0.33]	[0.71, 0.14]
Copeland		–	[1.00, 0.00]	[1.00, 0.00]	[0.20, 0.40]	[0.20, 0.40]	[-1.00, 1.00]	[1.00, 0.00]	[1.00, 0.00]
Dodgson			–	[1.00, 0.00]	[0.33, 0.33]	[-0.71, 0.86]	[-1.00, 1.00]	[0.71, 0.14]	[1.00, 0.00]
Max-min				–	[0.71, 0.14]	[-0.33, 0.67]	[-0.71, 0.86]	[0.33, 0.33]	[0.71, 0.14]
Plurality					–	[0.71, 0.14]	[0.33, 0.33]	[0.33, 0.33]	[-0.33, 0.67]
Borda						–	[0.33, 0.33]	[0.33, 0.33]	[-0.33, 0.67]
Approval							–	[-1.00, 1.00]	[-1.00, 1.00]
Runoff								–	[1.00, 0.00]
Nanson									–

Table 19 Values of the [degree of similarity, degree of dissimilarity] for the particular pairs of voting procedures using the Russell–Rao measures of similarity (20) and dissimilarity (21)

Voting procedure	Voting procedure								
	Amendment	Copeland	Dodgson	Max-min	Plurality	Borda	Approval	Runoff	Nanson
Amendment	–	[0.57, 0.43]	[0.29, 0.71]	[0.43, 0.57]	[0.29, 0.71]	[0.29, 0.71]	[0.14, 0.86]	[0.29, 0.71]	[0.43, 0.57]
Copeland		–	[0.43, 0.57]	[0.57, 0.43]	[0.43, 0.57]	[0.43, 0.57]	[0.14, 0.86]	[0.43, 0.57]	[0.57, 0.43]
Dodgson			–	[0.43, 0.57]	[0.29, 0.71]	[0.14, 0.86]	[0.00, 1.00]	[0.29, 0.71]	[0.43, 0.57]
Max-min				–	[0.43, 0.57]	[0.29, 0.71]	[0.14, 0.86]	[0.29, 0.71]	[0.43, 0.57]
Plurality					–	[0.43, 0.57]	[0.29, 0.71]	[0.29, 0.71]	[0.29, 0.71]
Borda						–	[0.29, 0.71]	[0.29, 0.71]	[0.29, 0.71]
Approval							–	[0.00, 1.00]	[0.00, 1.00]
Runoff								–	[0.43, 0.57]
Nanson									–

Table 20 Values of [degree of similarity, degree of dissimilarity] for the particular pairs of voting procedures using the Sokal–Michener measures of similarity (22) and dissimilarity (23)

Voting procedure	Voting procedure								
	Amendment	Copeland	Dodgson	Max-min	Plurality	Borda	Approval	Runoff	Nanson
Amendment	–	[0.86, 0.14]	[0.57, 0.43]	[0.71, 0.29]	[0.43, 0.57]	[0.43, 0.57]	[0.29, 0.71]	[0.57, 0.43]	[0.71, 0.29]
Copeland		–	[0.71, 0.29]	[0.86, 0.14]	[0.57, 0.43]	[0.57, 0.43]	[0.14, 0.86]	[0.71, 0.29]	[0.86, 0.14]
Dodgson			–	[0.86, 0.14]	[0.57, 0.43]	[0.29, 0.71]	[0.14, 0.86]	[0.71, 0.29]	[0.86, 0.14]
Max-min				–	[0.71, 0.29]	[0.43, 0.57]	[0.29, 0.71]	[0.57, 0.43]	[0.71, 0.29]
Plurality					–	[0.71, 0.29]	[0.57, 0.43]	[0.57, 0.43]	[0.43, 0.57]
Borda						–	[0.57, 0.43]	[0.57, 0.43]	[0.43, 0.57]
Approval							–	[0.14, 0.86]	[0.00, 1.00]
Runoff								–	[0.86, 0.14]
Nanson									–

Table 21 Values of [degree of similarity, degree of dissimilarity] for the particular pairs of voting procedures using the Rogers–Tanimoto measures of similarity (24) and dissimilarity (25)

Voting procedure	Voting procedure								
	Amendment	Copeland	Dodgson	Max-min	Plurality	Borda	Approval	Runoff	Nanson
Amendment	–	[0.75, 0.25]	[0.40, 0.60]	[0.56, 0.44]	[0.27, 0.73]	[0.27, 0.73]	[0.17, 0.83]	[0.40, 0.60]	[0.56, 0.44]
Copeland		–	[0.56, 0.44]	[0.75, 0.25]	[0.40, 0.60]	[0.40, 0.60]	[0.08, 0.92]	[0.56, 0.44]	[0.75, 0.25]
Dodgson			–	[0.75, 0.25]	[0.40, 0.60]	[0.17, 0.83]	[0.08, 0.92]	[0.56, 0.44]	[0.75, 0.25]
Max-min				–	[0.56, 0.44]	[0.27, 0.73]	[0.17, 0.83]	[0.40, 0.60]	[0.56, 0.44]
Plurality					–	[0.56, 0.44]	[0.40, 0.60]	[0.40, 0.60]	[0.27, 0.73]
Borda						–	[0.40, 0.60]	[0.40, 0.60]	[0.27, 0.73]
Approval							–	[0.08, 0.92]	[0.00, 1.00]
Runoff								–	[0.75, 0.25]
Nanson									–

Table 22 Values of [degree of similarity, degree of dissimilarity] for the particular pairs of voting procedures using the Kulczyński measures of similarity (26) and dissimilarity (27)

Voting procedure									
Voting procedure	Amendment	Copeland	Dodgson	Max-min	Plurality	Borda	Approval	Runoff	Nanson
Amendment	–	[4.00, 0.50]	[0.67, 0.80]	[1.50, 0.67]	[0.50, 0.82]	[0.50, 0.82]	[0.20, 0.92]	[0.67, 0.80]	[1.50, 0.67]
Copeland		–	[1.50, 0.67]	[4.00, 0.50]	[1.00, 0.70]	[1.00, 0.70]	[0.17, 0.92]	[1.50, 0.67]	[4.00, 0.50]
Dodgson			–	[3.00, 0.62]	[0.67, 0.80]	[0.20, 0.92]	[0.00, 1.00]	[1.00, 0.78]	[3.00, 0.62]
Max-min				–	[1.50, 0.67]	[0.50, 0.82]	[0.20, 0.92]	[0.67, 0.80]	[1.50, 0.67]
Plurality					–	[1.50, 0.67]	[0.67, 0.80]	[0.67, 0.80]	[0.50, 0.82]
Borda						–	[0.67, 0.80]	[0.67, 0.80]	[0.50, 0.82]
Approval							–	[0.00, 1.00]	[0.00, 1.00]
Runoff								–	[3.00, 0.62]
Nanson									–

(11)–(27), does confirm the very essence of results obtained by employing the more qualitative approach proposed in Sect. 3.

Namely, one can notice again that, not surprisingly, Copeland, Max-Min, Dodgson and Nanson form a group of voting procedures that have a high similarity and a low dissimilarity. Quite closely related to that group are Runoff and Amendment. By the way, except for Runoff, all these procedures are the Condorcet extensions, i.e. they result in the choice of the Condorcet winner if it exists. The so-called positional methods, that is, Plurality, Borda and Approval, seem to be rather far away from the rest of the procedures. This holds particularly for Approval. It can also be noticed that it is not very relevant which particular similarity and dissimilarity measure is actually used. The values obtained can be different but the order and proportions are maintained.

6 Concluding Remarks

We have presented a more comprehensive approach to a quantitative analysis of similarity and dissimilarity of voting procedures. We assumed a set of well known voting procedures and criteria which they should satisfy, which are known in political science (cf. Nurmi's [12] book). More specifically, we have considered the amendment, Copeland, Dodgson, max-min, plurality, Borda, approval, runoff, and Nanson, voting procedures, and the Condorcet winner, Condorcet loser, majority winner, monotonicity, weak Pareto winner, consistency, and heritage criteria. The satisfaction or dissatisfaction of the particular criteria by the particular voting procedures are represented as binary vectors. We used first rough sets to obtain a smaller number of voting procedures (9 instead of 13), following Fedrizzi, Kacprzyk and Nurmi [19], and then used the idea of Kacprzyk, Nurmi and Zadrożny [23] in which the use of some measures of similarity and dissimilarity for binary patterns has been proposed and the Jaccard–Needham measures have been used. In this paper we extend the above approach by using in addition to those, the similarity and dissimilarity measures of: Dice, Correlation, Yule, Russell–Rao, Sockal–Michener, Rodgers–Tanimoto, and Kulczyński.

References

1. Pitt, J., Kamara, L., Sergot, M., Artikis, A.: Voting in multi-agent systems. Comput. J. **49**(2), 156–170 (2006)
2. Pitt, J., Kamara, L., Sergot, M., Artikis, A.: Formalization of a voting protocol for virtual organizations. In: Dignum, F., Dignum, V., Koenig, S., Kraus, S., Singh, M., Wooldridge, M. (eds.) Proceedings of 4th AAMAS05, pp. 373–380. ACM (2005)
3. Pitt, J., Kamara, L., Sergot, M., Artikis, A.: Voting in online deliberative assemblies. In: Gardner, A., Sartor, G. (eds) Proceedings of 10th ICAIL, pp. 195204. ACM (2005)
4. Arrow, K.J., Sen, A.K., Suzumura, K (eds.): Handbook of Social Choice and Welfare, vol. 1. Elsevier (2002)

5. Kelly, J.S.: Social Choice Theory. Springer, Berlin (1988)
6. Plott, C.R.: Axiomatic social choice theory: an overview and interpretation. Am. J. Polit. Sci. **20**, 511–596 (1976)
7. Schwartz, T.: The Logic of Collective Choice. Columbia University Press, New York (1986)
8. Arrow, K.: A difficulty in the concept of social welfare. J. Polit. Econ. **58**(4), 328346 (1950)
9. Gibbard, A.: Manipulation of voting schemes: a general result. Econometrica **41**(4), 587601 (1973)
10. Kelly, J.S.: Arrow Impossibility Theorems. Academic Press, New York (1978)
11. May, K.: A set of independent, necessary and sufficient conditions for simple majority decision. Econometrica **20**(4), 680684 (1952)
12. Nurmi, H.: Comparing Voting Systems. D. Reidel, Dordrecht (1987)
13. Riker, W.H.: Liberalism against Populism. W. H. Freeman, San Francisco (1982)
14. Satterthwaite, M.A.: Strategy-proofness and arrows conditions: existence and correspondence theorems for voting procedures and social welfare functions. J. Econ. Theory, **10**, 187217 (1975)
15. Baigent, M.: Metric rationalisation of social choice functions according to principles of social choice. Math. Soc. Sci. **13**(1), 5965 (1987)
16. Elkind, E., Faliszewski, P., Slinko, A.: On the role of distances in defining voting rules. In: van der Hoek, W., Kaminka, G.A., Lesprance, Y., Luck, M., Sen, S., (eds.) Proceedings of 9th International Conference on Autonomous Agents and Multiagent Systems (AAMAS 2010), pp. 375–382 (2010)
17. McCabe-Dansted, J.C., Slinko, A.: Exploratory analysis of similarities between social choice rules. Group Decis. Negot. **15**(1), 77–107 (2006)
18. Richelson, J.: A comparative analysis of social choice functions I, II, III: a summary. Behav. Sci. **24**, 355 (1979)
19. Fedrizzi, M., Kacprzyk, J., Nurmi, H.: How different are social choice functions: a rough sets approach. Qual. Quant. **30**, 87–99 (1996)
20. Pawlak, Z.: Rough sets. Int. J. Inf. Comput. Sci. **11**, 341–356 (1982)
21. Pawlak, Z.: Rough Sets: Theoretical Aspects of Reasoning about Data. Kluwer, Dordrecht (1991)
22. Pawlak, Z., Skowron, A.: 2007a. Rudiments of rough sets. Inf. Sci. **177**(1), 3–27 (1988)
23. Kacprzyk, J., Nurmi, H., Zadrożny, S.: Towards a comprehensive similarity analysis of voting procedures using rough sets and similarity measures. In: Skowron, A., Suraj, Z. (eds.) Rough Sets and Intelligent Systems—Professor Zdzislaw Pawlak in Memoriam, vol. 1, pp. 359–380. Springer, Heidelberg and New York (2013)
24. Tubbs, J.D.: A note on binary template matching. Pattern Recogn. **22**(4), 359–365 (1989)
25. Choi, S.-S., Cha, S.-H., Tappert, ChC: A survey of binary similarity and distance measures. J. Syst. Cybern. Inf. **8**(1), 43–48 (2010)
26. Kacprzyk, J., Zadrożny, S.: Towards a general and unified characterization of individual and collective choice functions under fuzzy and nonfuzzy preferences and majority via the ordered weighted average operators. Int. J. Intell. Syst. **24**(1), 4–26 (2009)
27. Kacprzyk, J., Zadrożny, S.: Towards human consistent data driven decision support systems using verbalization of data mining results via linguistic data summaries. Bull. Pol. Acad. Sci., Tech. Sci. **58**(3), 359–370 (2010)
28. Yager, R.R.: On ordered weighted averaging operators in multicriteria decision making. IEEE Trans. Syst. Man Cybern. **SMC-18**, 183–190 (1988)
29. Yager, R.R., Kacprzyk, J. (eds.): The Ordered Weighted Averaging Operators: Theory and Applications. Kluwer, Boston (1997)
30. Yager, R.R., Kacprzyk, J., Beliakov, G. (eds): Recent Developments in the Ordered Weighted Averaging Operators: Theory and Practice. Springer (2011)
31. Polkowski, L.: A set theory for rough sets. Toward a formal calculus of vague statements. Fund. Inform. **71**(1), 49–61 (2006)
32. Pawlak, Z., Skowron, A.: Rough sets: some extensions. Inf. Sci. **177**(1), 28–40 (2007b)

33. Peters, J.F., Skowron, A., Stepaniuk, J.: Rough sets: foundations and perspectives. In: Meyers, R.A. (ed.) Encyclopedia of Complexity and Systems Science, pp. 7787–7797. Springer, Berlin (2009)
34. Greco, S., Matarazzo, B., Słowiński, R.: Rough sets theory for multicriteria decision analysis. Eur. J. Oper. Res. **129**(1), 1–47 (2001)
35. Pawlak, Z., Skowron, A.: Rough sets and Boolean reasoning. Inf. Sci. **177**, 41–73 (2007c)
36. Yao, Y.Y., Zhao, Y.: Discernibility matrix simplification for constructing attribute reducts. Inf. Sci. **179**, 867–882 (2009)
37. Straffin, P.D.: Topics in the Theory of Voting. Birkhäuser, Boston (1980)
38. Fishburn, P.C.: Condorcet social choice functions. SIAM J. Appl. Math. **33**, 469–489 (1977)

A Geo-Spatial Data Infrastructure for Flexible Discovery, Retrieval and Fusion of Scenario Maps in Preparedness of Emergency

Gloria Bordogna, Simone Sterlacchini, Paolo Arcaini,
Giacomo Cappellini, Mattia Cugini, Elisabetta Mangioni
and Chrysanthi Polyzoni

Abstract In order to effectively plan both preparedness and response to emergency situations it is necessary to access and analyse timely information on plausible scenarios of occurrence of ongoing events. Scenario maps representing the estimated susceptibility, hazard or risk of occurrence of an event on a territory are hardly generated real time. In fact the application of physical or statistical models using environmental parameters representing current dynamic conditions is time consuming on low cost hardware equipment. To cope with this practical issue we propose an off line generation of scenario maps under diversified environmental dynamic parameters, and a geo-Spatial Data Infrastructure (SDI) to allow people in charge of emergency preparedness and response activities to flexibly discover, retrieve, fuse and visualize the most plausible scenarios that may happen given some ongoing or forecasted dynamic conditions influencing the event. The novelty described in this chapter is related with both the ability to interpret flexible queries in order to retrieve risk scenario maps that are related to the current situation and to show the most plausible worst and best scenarios that may occur in each elementary area of the territory. Although, the SDI proposal has been conceived and designed to support the management of distinct natural and man-made risks, in the proof of concept prototypal implementation the scenarios maps target wild fire events.

Keywords Susceptibility · Hazard · Risk maps · Geo-Spatial data infrastructure · Database flexible querying · Decision support system

S. Sterlacchini · P. Arcaini · G. Cappellini · M. Cugini · E. Mangioni · C. Polyzoni
Istituto per la Dinamica dei Processi Ambientali, Consiglio Nazionale delle Ricerche
(CNR-IDPA), Piazza della Scienza 1, 20126 Milan, Italy
e-mail: simone.sterlacchini@idpa.cnr.com

G. Bordogna (✉)
Istituto per il Rilevamento Elettromagnetico dell'Ambiente, Consiglio Nazionale delle
Ricerche (CNR-IREA), via Corti 12, 20133 Milan, Italy
e-mail: bordogna.g@irea.cnr.it

© Springer International Publishing Switzerland 2017
J. Kacprzyk et al. (eds.), *Granular, Soft and Fuzzy Approaches
for Intelligent Systems*, Studies in Fuzziness and Soft Computing 344,
DOI 10.1007/978-3-319-40314-4_9

1   Introduction

The European Civil Protection Agency,[1] in its program regarding risk prevention, preparedness and emergency response states that the main activities for the reduction of impacts due to natural disasters are two: specifically, the improvement of methods of communication and information sharing between authorities, technicians, volunteers and the public, during the preparation of emergency response, and the increase of the level of information and education of the population that lives daily with the risks, in order to increase its awareness.

On the other side, the World Conference on Disaster Reduction (2005) has promoted initiatives to improve early warning systems aimed at identifying, estimating and monitoring risk levels, based on a direct involvement of the population (people-centered early warning systems): the main purpose of such initiatives is to increase the resilience of the community by both encouraging pro-active attitudes and reducing the potential damage.

Following these ideas, we have designed a geo-Spatial Data Infrastructure (SDI) with the ambition to increase the efficiency of preparedness and response activities to emergencies related to natural disasters carried out by local actors (authorities, civil protection, technicians and volunteers at the municipal level) having as main constraint the containment of the costs for the needed hardware and software equipment. We target local administrations, often small municipalities in Italy, located on the mountains, the most frequently touched by critic environmental crisis, which can hardly effort big investments. For such reasons we developed a flexible SDI by exploiting open source standard packages (namely, GrassGIS, PostgreSQL, Geonetwork, Ushahidi, and Geoserver) that can run on common PCs.

The basic core idea of the SDI is a catalogue service of susceptibility, hazard and risk scenario maps, that enables operators of the civil protection to discover the most plausible scenarios that may occur during an emergency management procedure. Since modelling susceptibility scenario maps, and consequently also hazard and risk maps depending on them, is a computationally time consuming activity with low cost hardware equipment, it is really difficult or even impossible to calculate them real-time when needed in order to face possible emergencies. In our case study, modelling a susceptibility map to wild fire at regional scale of an area of about 3000 square km can take up to half an hour on a common PC with a few gigabytes of memory. For this reason, in order to provide real-time information to the operators in charge of managing the pre-alarm/alarm phase, we decided to store many susceptibility maps, each one generated off-line with diversified conditions described by distinct values of the static and dynamic parameters that influence the occurrence of the phenomenon. These maps can be retrieved by the operator during an emergency preparedness or management phase by querying the catalogue service of the SDI. On the basis of these premises the SDI has been designed with several components connected via the Internet to perform the following tasks:

---

[1]http://ec.europa.eu/echo/civil_protection/civil/prote/cp14_en.htm.

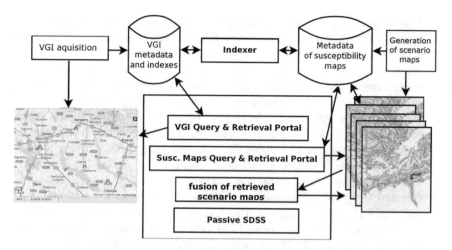

Fig. 1 Components of the geo-spatial data infrastructure for flexible discovery, retrieval and fusion of scenario maps during preparedness and emergency management

- *Generation of hazard and risk maps:* this off-line component (depicted in Fig. 1 as "generation of scenario maps") is made available to a pool of experts for the generation of susceptibility, hazard and risk scenario maps of the area being monitored. In our approach the modelling of forest fire (ignition and spread) and the identification of expected risk scenario maps is performed offline, usually in times of peace, and provides input information to the SDI system components. Distinct modelling techniques are applied depending on the type of phenomenon and on the representation scale [1, 2]. Each map shows areas of the monitored territory characterized by different values of probability of initiation and spread of the event based on the parameters used in the analysis phase. At the end of the modeling phase, each map is compared with the spatial distribution of the vulnerable elements (with the corresponding descriptors, including the economic value) in order to lead to the identification of a series of risk scenarios useful to provide information about the potential impact of direct/indirect related occurrence of the modelled event. This last phase may require the manual intervention of the expert.
- *Flexible discovery and retrieval of hazard and risk maps:* this component, the core of which is the catalogue service of the SDI (depicted in Fig. 1 comprising the "metadata of scenario maps", and the "susceptibility maps querying and retrieval portal") is made available to an operator of the civil protection to search, and retrieve real-time scenario maps stored in the geo-spatial database [3]. To allow the retrieval of the most plausible risk scenario maps that suit the current dynamic conditions, in an off-line procedure, metadata are associated to the scenario maps to represent the lineage of the maps, i.e., to describe the models and parameters used to generate them. Subsequently, an online procedure makes available a flexible query language to search the metadata. To model

flexible querying we rely on the framework of fuzzy databases [4–6]. The system searches and selects the "most plausible scenario maps", i.e. those that have been generated with parameter values that best meet the current soft constraints expressed in the flexible query by the operator. Then, it proposes either a ranked list of the stored scenarios maps ordered according to the degree of fulfilment of the soft constraints, or two virtual maps representing the best and worst scenarios that can be determined based on the fusion of the ranked retrieved maps [7, 8]. The pessimistic and optimistic plausible scenario maps are computed by applying an Induced Ordered Averaging Operator (IOWA) by the component named "fusion of retrieved scenario maps" depicted in Fig. 1 [9].

- *Volunteer Geographic Information Exploitation:* the SDI also manages Volunteered Geographic Information (VGI) (see VGI acquisition and VGI query and retrieval portal in Fig. 1). This information can be created by different categories of reporters (citizens, Civil Defence Volunteers, technicians, etc.) who may point out risky situations actually observed in a territory by connecting to an installation of the *Ushahidi* tool for VGI creation [10]. The SDI indexes the VGI reports and allows the operators to analyse them, to modify, to cancel or to publish them on the internet as open data [11].

- *Spatial Decision Support for Emergency Management:* this component, named "passive SDSS" in Fig. 1, is a passive Decision support System that can be activated by the operator of a municipality once an area in which either a dangerous situation has been reported or when a critical event in progress is identified. The SDSS allows guiding the operator through the flow of operations that the regulatory framework, laws in force at national and regional level, prescribe for that particular emergency situation. The regulation is formalized by a Petri net [12] which is executed step by step by the operator who can choose to perform only the actions in the process prescribed by the laws at each step, such as communicating with authorities, accessing resources to manage the situation (vehicles, equipment, etc.), directing on field operators to vulnerable and/ or strategic structures (hospitals, schools, industries, populated areas, etc.).

The innovation of this framework are several among which the ones we present in this chapter, that is, the off-line computation of several risk scenario maps by taking into account the influence of diversified dynamic parameters and their storage and content representation within a flexible SDI; the incorporation within an SDI for emergency management of a fuzzy database for flexibly querying metadata of scenarios maps; and finally the geo-spatial fusion mechanism that generates the best and worst scenario maps of the monitored territory as a result of a flexible query.

In the following sections we will describe, the procedure for the generation of hazard and risk scenario maps, the flexible query component, and the geo-spatial fusion mechanism.

2   Hazard and Risk Scenario Maps Generation

The aim of the hazard and risk scenario map generation component is to provide the end-users with several possible scenarios of occurrence of hazard and risk depending on different conditions of the parameters that influence the modelled phenomenon. Preliminary to the generation of hazard and risk scenario maps is the generation of probability-based susceptibility maps. In our case study we modelled wildfire susceptibility maps at a regional scale, considering different types of geo-environmental and meteo-climatic variables which are the conditions influencing wildfires outbreak and spreading. The study area is located in the central-eastern part of the Sardinia Island (Nuoro Province, Italy) with a geographical extension of about 3.934 km². Data concerning the triggering points, temporal occurrence of events and geo-environmental variables have been downloaded from the Nuoro Province geoportal (www.sardegnageoportale.it/catalogodati/website); meteorological parameters have been made available by Epson-Meteo, an Italian weather fore-casting agency.

In this study, the hazard modeling procedure has not been completely performed: a hazard map should classify the study area according to different wildfire spatial occurrence probabilities (concerning the triggering points and the distance and direction of propagation of the fire), temporal occurrence probabilities (return periods) and the magnitude of the expected events. Given the regional scale adopted in this study, the final maps only provide the end-users with the spatial probability of the expected damaging events; no information concerning the magnitude and the temporal probability of the events is made available (although this information has been derived qualitatively by analyzing past events). For this reason, such maps are indeed wildfire susceptibility maps and allow identifying areas characterized by different proneness to wildfire in terms of ignition (triggering points) and propa-gation. This information is crucial during the emergency preparedness phase to let the disaster manager be aware of the expected number of people that could be affected by fire propagation and the expected level of damage to infrastructure.

In Fig. 2 the process of generation of susceptibility maps is depicted.

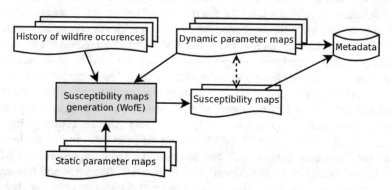

Fig. 2 Generation of susceptibility maps by the supervised learning method (Weight oF Evidence)

Concerning the modeling techniques applied in this study, a data-driven Bayesian method, Weights of Evidence (WofE) has been successfully applied to identify the most susceptible areas to ignition [1, 2].

WofE is the log-linear form of the well-known data-driven Bayesian learning method; it uses known wildfire occurrences as training points and evidential themes, representing explanatory variables, to calculate prior (unconditional) and posterior (conditional) probabilities. The prior probability is the probability that a pixel contains a wildfire occurrence considering only the number of the training points, and not taking into account neither their location nor any evidential themes available for the study. On the contrary, the posterior probability is the probability that a pixel contains a wildfire occurrence based on explanatory variables, provided by the evidential themes. WofE calculates the degree of spatial association among training points and each explanatory variable class by means of positive (W+) and negative (W−) weights [13]. A positive correlation shows W+ positive and W − negative; a negative correlation is represented by W− positive and W+ negative.

W+ = W− = 0 calls for a lack of correlation between the wildfire occurrence and the explanatory variable class and in such case the posterior probability equals the prior probability.

Concerning the triggering points in the study area of Nuoro province in Sardinia island, all past events since 2005–2012 have been used. Each triggering point is described spatially (in terms of its geographical position) and temporally (in terms of day, month and year of occurrence). Then, a wildfire inventory has been compiled by identifying a total number of 1.698 events: only those whose area is less than 2 ha have been considered in the analysis (about 60 % of the total burnt area) and represented by using a single point. After that, the 986 ignition points have been randomly subdivided into two mutually exclusive subsets (each containing 50 % of the total number of wildfires): the success subset (training points set) and the prediction subset (test set). The former (including 493 initiation points) has been used to calibrate the model while the latter (including 493 initiation points) to validate the model. These subsets have allowed success and prediction rate curves to be computed for assessing the robustness of the model [2, 13, 14]. By computing the Success Rate Curve (SRC) we estimated how much the results of the models fit the occurrence of wildfires used for the training of the models (success subset) in each experiment, while the Prediction Rate Curve (PRC) was computed to validate the models since it estimates how much the model correctly predicts the occurrence of wild fires in the validation set (prediction subset) not used in the experiments.

Explanatory variables have been subdivided into two broad categories: static and dynamic [15]: the former comprises variables that could be considered invariant within the entire period of the study: altitude, slope, aspect, land use, vegetation, distance from roads, population density. The latter, on the contrary, includes parameters that could change in time following different rules from zone to zone: wind speed and direction; rainfall and temperature. Grid maps concerning temperature and rainfall refer to different periods: one-day, one-week and one-month before the day of occurrence. For wind speed and directions, only one-day grid

Table 1 Table presents the best results in terms of PSR and PPR values; they refer to the 30 % of the most susceptible area. The PSR value computed for static and dynamic variables refers specifically to the month of August, one of the most affected by wildfire; the PPR value has not been assessed for the model using both static and dynamic variables due the low number of triggering points detected in the period

	Using only static variables	Using both static variables and dynamic variables
PSR	75.7	81.5
PPR	70.6	–

maps before the date of occurrence were available. All explanatory variables have been tasselled by using a pixel size of 25 m^2.

The WofE has been applied several times by using the Spatial Data Modeller [16] and by changing the combination of explanatory variables so as to produce several susceptibility maps. First, only static parameters have been analysed; after that, the role of meteo-climatic variables on the final wildfire "static" susceptibility maps have been assessed by including them in the analysis.

Since in the case of susceptibility prediction it is important to evaluate the ability of the model to detect wildfire occurred in reality with respect to all its positive detections, the Positive Success Rate values (PSR) and Positive Prediction Rate values (PPR) shown in Table 1 have been computed.

It can be seen that the PSR increases by considering the dynamic variables in the modelling procedure.

The results of a modelling phase is a wildfire susceptibility map that represents the spatial domain under analysis by probability values on a continuous scale [0, 1].

In our case study, at the end of the modelling phase, a collection of 100 wildfire susceptibility maps are available, each of which modelled considering different subsets of variables and constraints. These maps are the scenarios maps. The information on the explanatory values used and modelling parameters are stored in metadata so that different civil protection operators may query and retrieve maps on the basis of relevant criteria mainly consisting in observed/measured values concerning both static geo-environmental variables and dynamic meteorological parameters. This operation can be performed both in preparedness and emergency response phases.

The results of this first part of the analysis is the main input of the following step concerning the identification of areas characterized by the presence of vulnerable elements that are located in the most wildfire susceptible zones or along the most probable fire directions. Each risk scenario will provide the end-users with an exposure value expressed in monetary terms and concerning the market values for public and private buildings or the reconstruction costs for infrastructures. In this way, in the framework of a long-term spatial planning overview, a list of priorities of intervention can be drawn up on the basis of the economic values exposed and the number of people living in each risk scenario.

3 Representing and Discovering Scenario Maps

This section describes the components of the SDI depicted in Fig. 3 devoted to search the scenario maps which in our case study are wild fire susceptibility maps.

In order to implement a discovery mechanism able to retrieve scenario maps plausibly depicting the current scenario, it is necessary to represent the contents of the parameter maps (that have been used to generate the scenarios maps) to empower the discovery facility of the SDI. To this aim a database of metadata is generated which contains, for each map, metadata records that synthesize the map lineage, i.e., the contents of the associated parameter maps. More specifically, each susceptibility map has been generated by a model using given parameter maps, in which the values of the parameters specify the static and dynamic conditions in each distinct position of the territory. We represent the content of such parameter maps by metadata and organize them into a database made available for querying and possibly retrieving the scenario map(s) which are the most "plausible" for the current situation, in relation to the environmental current conditions, i.e., current parameters. We distinguish between dynamic parameters that change over time, such as those denoting meteorological conditions, and static parameters that, given a geographic area, can be considered invariable over the time of analysis, such as land use, vegetation, altitude, slope aspect and gradient, among the others. In the case study, the dynamic parameters are the temperature and rainfall, with reference to the day before the occurrence, one week before, and one month before; and wind direction and speed, with reference to the day before the occurrence.

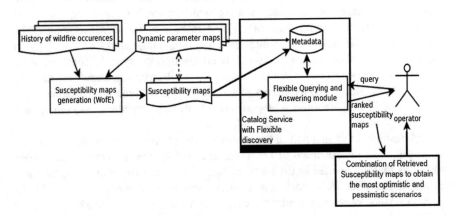

Fig. 3 Metadata generation and discovery of scenario maps

3.1   Metadata Definition

Each scenario map is represented by a metadata record that roughly synthesises its content:

$$< sm, SM, N, M, D, [PN], Hs > \tag{1}$$

where:

- sm is the name of the type of hazard/susceptibility we are considering (in the case study it is equal to "wild fire susceptibility");
- SM is the unique identifier of the Scenario Map;
- N and M are the dimensions in pixels of the map;
- D is the domain of values of SM (for example susceptibility values are defined in $[0,1]$),
- $[PN]$ is a vector of unique identifiers of Parameter maps names used to generate SM;
- $Hs = \{< vi, \phi_{vi} > , < v2, \phi_{v2} > ,..., < vn, \phi_{vn} >\}$ is a level-2 fuzzy set [17], where the linguistic values $v1, .., vn$ are ordered based on their trapezoidal-shaped membership functions on D: $\forall i < j$ $\mathrm{argmax}_{x \in D}(\mu_{vi}(x)) < \mathrm{argmax}_{x \in D}(\mu_{vj}(x))$. Each membership degrees $\phi_{vk} \in [0,1]$ is defined as the relative fuzzy cardinality of vk over the values of the pixels in the map SM:

$$phi_{vk} = \frac{\sum_{i,j}^{N,M} \mu_{vk}(SM_{ij})}{N*M \max_{i,j=1,..N; 1,...M}(SM_{ij})}. \tag{2}$$

Thus Hs can be regarded as a representation of the quantized normalized histogram of frequency of the values in SM, according to a fuzzy partition defined by the linguistic values $v1, v2,..., vn$ on D.

Let us make an example of Hs computation on the synthetic scenario map SM depicted in Table 2, by considering the quantisation defined by the following three linguistic values defined by trapezoidal membership values:

$v1 = low$ with $\mu_{low} = (0, 0, 0, 0.5)$
$v2 = medium$ with $\mu_{medium} = (0, 0.5, 0.5, 1)$
$v3 = high$ with $\mu_{high} = (0.5, 1, 1, 1)$

where the quadruples $(0, 0, 0, 0.5)$ $(0, 0.5, 0.5, 1)$ and $(0.5, 1, 1, 1)$ define the vertexes of the triangles depicted in Fig. 4.

Table 2 Example of scenario map SM of dimensions $N = 4$, $M = 4$ and domain $D = [0, 1]$						
1				1	0.25	0.5
1				1	0.25	0.5
1				1	0	0
1				0	0	0

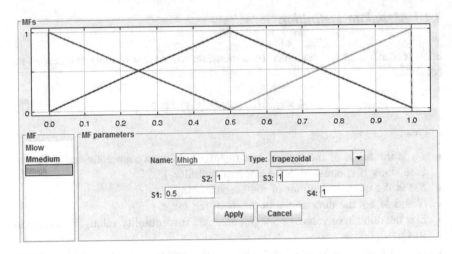

Fig. 4 Membership functions defined by the quadruples $\mu_{low} = (0, 0, 0, 0.5)$, $\mu_{medium} = (0, 0.5, 0.5, 1)$ $\mu_{high} = (0.5, 1, 1, 1)$

We obtain the following membership degrees to HS:

$\phi_{low} = (\Sigma_{i=1,\ldots,4;\ j=1,\ldots,4}\ \mu_{low}(SM_{ij})/16 = 6/16$
$\phi_{medium} = 3/16$
$\phi_{high} = 7/16$
and thus $Hs = \{< low, 0.38 >, < medium, 0.19 >, < high, 0.44 >\}$.

With this representation we can guess that the scenario map SM contains primarily *high* and *low* values and marginally *medium* values.

Furthermore, each scenario map SM has associated a number of dynamic parameter maps PN uniquely represented by a metadata record that roughly synthesizes the parameter map content. This metadata record is structured into the following fields:

$$< pn, PN, TP, [SM], N, M, D, Hp, D', Hp' > \qquad (3)$$

where:

- pn is the parameter name;
- PN is the unique identifier of the name of the parameter map;
- $TP \in \{last\ month,\ last\ week,\ yesterday\}$ is the time period of reference of the parameter PN;
- $[SM]$ is a vector containing the unique identifiers of the SM maps that were generated with the contribution of parameter PN;
- N and M are the dimensions of PN;
- D is the domain of values of the parameter (for example integer);
- $Hp = \{<v1,\phi_{v1} >, < v2,\phi_{v2} >,\ldots, < vn,\phi_{vn} >\}$ is a level 2 fuzzy set defined as Hs above. The linguistic values $v1, \ldots, vn$ are ordered by their trapezoidal-shaped

membership functions on D, i.e., $\forall i < j$ $\mathrm{argmax} x \in D(\mu_{vi}(x)) < \mathrm{argmax}_{x \in D}$ $(\mu_{vj}(x))$. The membership degrees $\phi_{vk} \in [0,1]$ are defined as the relative fuzzy cardinality of vk over the pixel values in the parameter map PN:

$$\phi_{vk} = \frac{\sum_{i,j}^{N,M} \mu_{vk}(PN_{ij})}{N*M \max_{i,j=1,..N; 1,...M}(PN_{ij})} \qquad (4)$$

Thus Hp is a representation of the quantised normalized histogram of frequency of the values in map PN according to a fuzzy partition $v1, v2,..., vn$ defined on D.

- D' is the domain of values of the parameter first derivative.
- $Hp' = = \{<dv1,\phi_{dv1}>, <dv2,\phi_{dv2}>,...,<dvn,\phi_{dvn}>\}$ is a level 2 fuzzy set defined as Hs and Hp above. The linguistic values $dv1, .., dvn$ are ordered by their trapezoidal-shaped membership functions on D', $\forall i < j$ $\mathrm{argmax}_{x \in D'}$ $(\mu_{dvi}(x)) < \mathrm{argmax}_{x \in D'}(\mu_{dvj}(x))$. The membership degrees $\phi_{dvk} \in [0,1]$ are defined as the relative fuzzy cardinality of dvk over the pixel values in the first derivate of the parameter map PN. Thus the values dvi define approximate ranges of the gradient map PN' of PN. The membership degrees $\phi_{dvk} \in [0, 1]$ are defined as the relative fuzzy cardinality of dvk over the first derivate of pixel values in the parameter map PN:

$$phi_{dvk} = \frac{\sum_{i,j}^{N',M'} \mu_{dvk}(PN'_{ij})}{N'*M' \max_{i,j=1,..N'; 1,...M'}(PN'_{ij})} \qquad (5)$$

3.2 Storing the Metadata in a Catalogue Service

The metadata records described above are embedded within an XML metadata format structured into fields compliant with current regulations in force at European level for geo-data interoperability representation [18]. Thus they determine an extension of the standard INSPIRE metadata format, guarantying full compliance with OGC Web Catalogue Services as far as the discovery facility based on spatial, time and semantic representation. The metadata records in (1), describing the scenario map content, are embedded within the content field of the standard metadata format, since they enable a content-based discovery of the scenario maps as described below, while the metadata records in (3), describing the associated parameter maps, are embedded within the lineage metadata field since they represent information on the parameters of the process that generated the scenario maps.

3.3 Discovering Scenario Maps

The Discovery of scenario maps is performed by querying the metadata catalogue service, which is a web application enabling online querying of the database of metadata, and together with the metadata retrieves the link of the associated scenario map. The user query is evaluated against the metadata indexes and both the metadata that satisfy it and the associated scenario maps are retrieved. Generally, current catalogue services are implemented on top of classic database management systems, such as *PostgreSQL*, no ranking based on relevance evaluation of the scenario maps is performed since the query language, basically *SQL*, allows expressing only crisp selection conditions. Nevertheless, in the context of emergency management, the possibility to retrieve an ordered list of scenario maps in decreasing order of plausibility to represent the current scenario is very important.

Typically, the user of the catalogue service is an operator of either the municipality or the local civil protection who is preparing the activities preliminary in view of a possible occurrence of an emergency, and is monitoring an ongoing critical situation.

He needs to retrieve the scenario maps described by the dynamic influencing parameter close to the current ones or similar to the forecasted ones, such as rainfalls and temperature values close either to those of last days, or to those forecasted for the next hours. Such scenario maps are likely to depict the current ongoing territorial situation, or the situation that may occur in the immediate future. The scenarios maps can aid operators to take decisions such as in planning evacuation of the population residing in areas at high risk, in identifying the safe areas where to collect the evacuated citizens, in organizing rescue operations. Several kinds of flexible queries can be formulated by the operators to take better decisions. To model flexible querying, we rely on the framework of fuzzy databases [4–6].

3.4 Direct Query by Constraining the Parameters Values

During an emergency operation a flexible direct query can specify soft conditions Sc on the values of some dynamic parameter pn: Sc: $D_{PN\_TP} \rightarrow [0, 1]$ where D_{PN_TP} is the domain of parameter named pn for the time period TP.

For example, one can specify a query composed by ANDing the following soft conditions:

high pn = temperature during TP = last month AND
very high pn = temperature during TP = last week AND
very high pn = wind during TP = yesterday AND
None pn = rainfall during TP = last month

The operator formulates this query when interested in the most plausible scenario maps given the current situation described by the linguistic values of the parameters specified in the soft conditions. Such soft conditions constrain the values of the correspondent parameter map *PN* to some ongoing recent climatic conditions, expressed by the linguistic values *high* for *temperature* during the *last month*, *very high* for *temperature* during the *last week*, etc.

Another situation in which this query can be formulated is when the operator has some forecasted values of the parameters and wants to obtain the most plausible scenario based on the forecasts.

This query is evaluated by accessing the metadata records indexed by both *pn* and *TP* and by matching the *Hp* field of each parameter map *PN* with the soft condition *Sc* to compute the ranking score r_{Sc} as follows:

- firstly, the compatibility of the query soft constraint *Sc* with the meaning μ_{vk} of a linguistic value *vk* in *Hp* is computed as the *min*;
- this compatibility degree is then aggregated by the *min* with the degree ϕ_k associated with *vk* in *Hp*:

$$r_{Sc} = \min_{k \in Hp} \left(\min_{i \in D_{PN\_TP}} (\mu_{vk}(i), Sc(i)), \phi_k) \right) \qquad (6)$$

This is done for all the soft conditions in the query, and finally the query ranking scores r_q is obtained by the minimum of r_{Sc}, since in the query the soft conditions are ANDed.

The scenario maps with ranking scores greater than zero are listed in decreasing order to the operator as shown in Fig. 5.

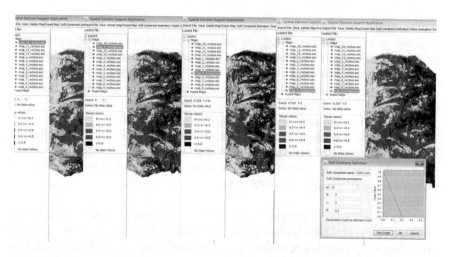

Fig. 5 Scenario maps ranked from left to right with respect to their degrees of satisfaction of the soft constraint *very low* depicted in the *right low corner*

3.5   Direct Query by Constraining the Parameters Trends

The operator specifies a soft condition on the trend of some parameters *pn* based on either recent values or forecasted values: $St: D'_{PN\_TP} \rightarrow [0,1]$ where D'_{PN_TP} is the domain of the first derivate of parameter *pn* during period *TP*. An example of such query can be:

High_increasing pn = temperature during TP = last month AND
High_increasing pn = wind during TP = yesterday AND
High_decreasing pn = rainfall during TP = last month

This query is evaluated by accessing the metadata records indexed by both *pn* and *TP* and by matching *Hp'* values *dvk* with the soft condition *Sc'* to compute the ranking score r_{Sc}' as in the previous case:

$$r'_{Sc} = \min_{k \in Hp'} (\min_{i \in D'_{PNderscoreTP}} (\mu_{vk}(i), Sc'(i)), \phi_k) \tag{7}$$

This is done for all the soft conditions in the query, and finally the query ranking scores are combined by the minimum again.

The scenario maps *SM* associated with the positive ranking scores to the query can be listed in decreasing order to the operator.

3.6   Inverse Query

Inverse queries are useful for territorial management planning, when a risk management team, responsible of a given territorial area, determines the lower/upper bounds of some environmental dynamic parameters *pn* which cause a given risk level. This information can be used to define pre-alert emergency plans in case of meteorological conditions within the lower and upper bounds. In this case the result of a query is not a scenario map but a summary of the parameter maps that generated the scenario map best satisfying the query condition.

The operator specifies a soft condition on the domain of *sm* with the aim of retrieving the values of the parameters *pn* that more likely determine the susceptibility values satisfying the soft constraint in most of the areas of the maps *SM*. An example of query could be the following:

   high sm

where *high* is a soft condition on the values of the scenario maps *SM* of type *sm*.

This query is evaluated by first defining the soft condition $Sc: D_{SM} \rightarrow [0, 1]$ and evaluating it by applying formula (6) on *SM* maps. Then the *SM* map with best matching score is selected: $SM \mid r_{Sc}(SM) = \text{ArgMax}(r_{Sc}(SM_1), \ldots, r_{Sc}(SM_v))$.

All the correspondent fuzzy sets *Hp* which synthetize the contents of the parameter maps *PM* associated with the selected scenario map *SM* are selected and shown to the operator.

3.7 Scenario Map Fusion

The result of a flexible query must be an informative answer. Besides retrieving a ranked list of the most plausible susceptibility maps, two virtual maps, representing the most likely optimistic and pessimistic scenarios that may occur, together with the variability of susceptibility levels in each pixel, are generated. To provide this additional answer, we perform a spatial fusion of the ranked retrieved maps.

Specifically, given M scenario maps retrieved by a query, we fuse them pixel by pixel and by taking into account their rank in order to obtain the most plausible optimistic and pessimistic scenario maps.

The plausible optimistic scenario map is the one in which the value of a pixel is computed based on the fusion of most of the smallest co-referred pixel values appearing in the top ranked retrieved maps. This is because the greater the pixel value the greater the susceptibility or risk, and then fusing the smallest values means being optimistic on the scenario that may happen.

In fact, in the optimistic map, we expect that the best case happens, i.e., we fuse the lowest susceptibility or risk levels within each pixel.

Conversely, for the plausible pessimistic scenario, we expect the worst situation to occur, i.e., the pessimistic scenario map is the one in which the value of a pixel is computed based on the fusion of most of the greatest co-referred pixel values in the top ranked retrieved maps. This is because the greater the pixel value the greater the susceptibility or risk and then fusing the greatest values means being pessimistic on the scenario that may happen.

The generation of the virtual scenario is performed by applying a pixel based fusion taking into account the ranks of the retrieved scenario maps, and an aggregation defined by an IOWA operator [9].

3.7.1 Definition of the IOWA Operator

The *IOWA* operator is defined in order to allow associating with each element to aggregate a priority defined by an external vector $(u_1,..., u_M)$, named inducing order vector. The *IOWA* operator of dimension M is a non linear aggregation operator $IOWA:[0,1]^M \rightarrow [0,1]$ with a weighting vector $W = (w_1, w_2, ..., w_M)$, with $w_j \in [0,1]$ and $\Sigma_{i=1,..M} w_i = 1$ defined as [9]:

$$IOWA(< x_1, u_1 > , ..., < x_M, u_M > = \Sigma_{i=1,..M} w_i * x_{u-index(i)} \qquad (8)$$

in which $X = (x_1,..., x_M)$ is the argument vector to be aggregated and $x_{u-index(i)}$ is the element of vector X associated with the i-th greatest inducing order value u among the values $(u_1,..., u_M)$.

For computing the plausible optimistic and pessimistic scenario maps, the inducing order vectors U^P and U^O are defined by taking into account both the pixels susceptibility degrees x_j, and the ranks r_j, of the scenarios maps in the retrieved list ($List_q$) as follows:

$$u_i^P = \frac{x_i}{max_{x_k \in SM} x_k} * \frac{|List_q| - r_i + 1}{|List_q|} \qquad u_i^O = \frac{max_{x_k \in SM} x_k - x_i}{max_{x_k \in SM} x_k} * \frac{|List_q| - r_i + 1}{|List_q|} \qquad (9)$$

In which $|List_q|$ is the cardinality of the retrieved list by query q. We could even choose to consider the first top K elements in the list $|List_q| = K$.

The weighting vector W of the IOWA operator is defined with the weights $w_i > 0$ $\forall i < k < |List_q|$ and $w_i = 0$ for $i \geq k$. This way the result is computed by the weighted combination of the top values x_i ordered in decreasing value of either U^P and U^O to generate the most plausible pessimistic and optimistic scenarios respectively.

3.7.2 Numeric Example

Let us consider the following values of the susceptibility degrees defined in [0,100]:

$X = (10\ 20\ 100\ 70\ 100\ 20)$ and their associated ranks $r = (1\ 2\ 3\ 4\ 5\ 6)$ in the retrieved list $List_q$, by applying formula (9) we compute the following induced order vectors U^P and U^O for the pessimistic and optimistic fusion:

$$U^P = (0.1\ 0.2\ 1\ 0.7\ 1\ 0.2)(1\ 0.8\ 0.6\ 0.5\ 0.2\ 0.1)^T. = (0.1\ 0.16\ 0.6\ 0.35\ 0.2\ 0.02)$$
$$U^O = (0.9\ 0.8\ 0\ 0.3\ 0\ 0.8)\ (1\ 0.8\ 0.6\ 0.5\ 0.2\ 0.1)^T. = (0.9\ 0.64\ 0\ 0.15\ 0\ 0.08)$$

Consequently the values of the susceptibility degrees X, reordered according to U^P and U^O respectively are the following:

$$X^P = (\ 100 \quad 70 \quad 100 \quad 20 \quad 10 \quad 20\ )$$
$$X^O = (\ 10 \quad 20 \quad 70 \quad 20 \quad 100 \quad 100\ )$$

Now let us consider the IOWA operator with weighting vector W defined such that $w_i = 0.33\ \forall i < 3$, and $w_i = 0$ otherwise. When we apply it to X with recorder vector U^P and U^P the following results are obtained:

- the most plausible pessimistic susceptibility degree is computed as follows: $X^P *$ $W = (100\ 70\ 100\ 20\ 10\ 20)\ (0.33\ 0.33\ 0.33\ 0\ 0\ 0)^T = 89$
- the most plausible optimistic susceptibility degree is computed as follows: $X^O *$ $W = (10\ 20\ 70\ 20\ 100\ 100)\ (0.33\ 0.33\ 0.33\ 0\ 0\ 0)^T = 33$

4 Flexible Spatial Data Infrastructure Implementation

Since the early 1990s, SDIs have been introduced to increase the availability and accessibility of geographic information on the Web. The SDI concept is focused on distributed geo-data and processes, and fundamental operations such as discovering, viewing, accessing, and integrating geo information. These operations are made possible thanks to the interoperability of web services, acting together through well-defined protocols and interfaces.

Interoperability is therefore a cornerstone in modern SDIs. Distributed and heterogeneous geospatial data managed at their web sites can be discovered by users who access on the Internet and query catalogue services managing metadata, i.e., semi-structured textual descriptions of the characteristics of distributed geo-data. Furthermore, such services provide access to the geo-data deemed relevant by a user after a discovery phase without copying or transferring any dataset. SDIs are therefore particularly attractive when geo-information sources are distributed and heterogeneous.

The geo SDA presented in this chapter has been implemented based on open source packages that have been extended in order to evaluate flexible queries and to return fused maps. To manage and deploy on the web the scenario maps we used *Geoserver*, that is a Web GIS compliant with the OGC standard services. All the scenario maps are served by *Web Map Service*.

For both the metadata management, querying and retrieval we used *Geonetwork* that is a package developed by *FAO* on top of a geographic database implemented by *PostgreSQL*, extended by *PostGIS* for managing the geometric attributes of the geographic bounding boxes. This object-relational database has been extended in order to evaluate flexible queries specifying soft conditions in the where clause of the basic *SQL* query.

The Fusion module has been implemented on top of *Openlayer* library and is called by *Geonetwork* anytime a user asks to display the most plausible optimistic or pessimistic scenario map satisfying a query.

5 Conclusions

Public territorial planning administrations need up-to-date geo-located information relative to the knowledge of critical situations that are happening or that are likely to happen in the near future under some dynamic conditions (both meteorological and environmental) in order to effectively plan and perform mitigation and emergency interventions.

This is the motivation of our proposal to design and develop a geo SDI allowing to access geo-located information such as authoritative information on the resources, vulnerable structures and contact people to be involved in the interventions, VGI relative to critical situations freely provided by testimonies of critical events, and finally scientific information on the possible scenarios of risk that may plausibly happen in a region given the occurrence of meteorological and environmental conditions. To this end we applied methods of fuzzy database and fuzzy data fusion to flexibly query the metadata of scenario maps and to generate virtual scenario maps that depict the most plausible optimistic and pessimistic situation that may possibly occur given the dynamic conditions specified in a query.

Although several projects generate susceptibility maps for forecasting events, they usually only permit to display the susceptibility maps and to do standard geospatial data manipulation operations (e.g., zooming and panning), but do not support flexible querying on maps contents nor map fusion. Some applications have modelled environmental risks based on dynamic variables such as in [19] where malaria disease is modelled by considering the climatic changes, and in [15], where they use fire simulators to predict fires, while the history of fires is only used to calibrate the input variables of the simulator (using genetic algorithms).

As far as we know, no approach has proposed a geo SDI for flexible querying stored scenarios maps, generated by models which consider the contributions of the dynamic parameters that have an influence on the modelled phenomena.

To summarize our proposed geo SDI is original for several aspects.

First of all the way in which we synthetically represent the contents of susceptibility scenario maps, and parameters maps, based on level-2 fuzzy sets and the extension of INSPIRE metadata structure so as to be able to empower the catalogue service with the ability to answer content-based queries on scenario maps.

Second, by adopting a fuzzy database approach, the catalogue service can rank the metadata, and thus the retrieved scenario maps described by them, with respect to flexible queries, providing indications on the most plausible scenarios that may occur based on the conditions in the query.

Finally, the pixel based fusion function allows computing two virtual scenario maps, synthesising the most plausible optimistic and pessimistic scenarios that may occur given the soft conditions described by the flexible query.

Aknowledgements This work has been carried out within the project SISTEMATI—Strumenti Informatici per lo Studio e il Trattamento di Emergenze Ambientali, Tecnologiche e Infrastrutturali and SIMULATOR—Sistema Modulare per la prevenzione dei rischi. Both projects have been funded by Regione Lombardia.

References

1. Bonham-Carter, G.F., Agterberg, F.P., Wright, D.F.: Integration of geological datasets for gold exploration in Nova Scotia. Photogram. Eng. **54**, 1585–1592 (1988)
2. Bonham-Carter, G.F., Agterberg, F.P., Wright, D.F.: Statistical pattern integration for mineral exploration. In: Gaal, G., Merriam, D.F. (eds.) Computer Applications in Resource Estimation: Prediction and Assessment for Metals and Petroleum, pp. 1–21. Pergamon Press, Oxford (1990)
3. Arcaini, P., Bordogna, G., Sterlacchini, S.: Wildfire susceptibility maps flexible querying and answering. In: Proceedings of FQAS 2013. Springer LNCS series, pp. 352–363 (2013)

4. Bordogna, G., Psaila, G.: Fuzzy-spatial SQL. In: Flexible Query Answering Systems. LNAI, vol. 3055, pp. 307–319. Springer (2004)

5. Bosc, P., Kaeprzyk, J.: Fuzziness in Database Management Systems. Studies in Fuzziness and Soft Computing, vol. 5. Physica-Verlag (1996)

6. Galindo, J.: Handbook of Research on Fuzzy Information Processing in Databases. IGI Global (2008)

7. Bordogna, G., Sterlacchini, S.: A multi criteria group decision making process based on the soft fusion of coherent evaluations of spatial alternatives. In: Proceedings of the 2nd World Conference on Soft Computing, BAKU, Azerbaijan, 3–5 Dec 2012

8. Pasi, G., Yager, R.: Modeling the concept of majority opinion in group decision making. Inf. Sci. **176**, 390–414 (2008)

9. Yager, R., Filev, D.P.: Induced ordered weighted averaging operators. IEEE Trans. Syst. Man Cybern. Part B. **29**(2), 141–150 (1999)

10. Okolloh, O.: Ushahidi, or 'testimony' : Web2. 0 tools for crowd sourcing crisis information. Particip. Learn. Action **59**(1), 65–70 (2009)

11. Arcaini, P., Bordogna, G., Sterlacchini, S.: Flexible querying of volunteered geographic information for risk management. In: Proceedings of the 8th Conference of the European Society for Fuzzy Logic and Technology (EUSFLAT 2013), Milano, Italy (2013)

12. Murata, T.: Petri nets: properties, analysis and applications. Proc. IEEE **77**, 541–580 (1989)

13. Raines, G.: Evaluation of weights of evidence to predict epithermal-gold deposits in the great basin of the Western United States. Nat. Resour. Res. **8**(4), 257–276 (1999)

14. Agterberg, F.P., Bonham-Carter, G.F., Wright, D.F.: Weights of evidence modelling: a new approach to mapping mineral potential. In: Agterberg, F.P., Bonham-Carter, G.F. (eds.) Statistical Applications in the Earth Sciences. Geological Survey of Canada, pp. 171–183 (1989)

15. Wendt, K.: Efficient knowledge retrieval to calibrate input variables in forest fire prediction. Master's thesis, Escola Tècnica Superior d'Enginyeria. Universitat Autònoma de Barcelona (2008)

16. Sawatzky, D.L., Raines, G.L., Bonham-Carter, G.F., Looney, C.G.: Spatial Data Modeller (SDM): ArcMAP 9.2 geoprocessing tools for spatial data modelling using weights of evidence, logistic regression, fuzzy logic and neural networks (2008). (http://arcscripts. esri.com/details.asp?dbid=15341)

17. Dubois, D., Prade, H.: Fuzzy Sets and Systems: Theory and Applications. Mathematics in Science and Engineering, vol. 114, pp. 50–63. Academic Press (1980)

18. Commission Regulation (EC) No 1205/2008 of 3 December 2008 Implementing Directive 2007/2/EC of the European Parliament and of the Council as regards metadata (2008)

19. Craig, M., Snow, R., le Sueur, D.: A climate-based distribution model of malaria transmission in Sub-Saharan Africa. Parasitol. Today **15**(3), 105–111 (1999)

The Multiple Facets of Fuzzy Controllers: *Look-up-Tables—A Special Class of Fuzzy Controllers*

Dimitar Filev and Hao Ying

Abstract Look-up table (LUT) controllers are among the most widely utilized control tools in engineering practice. The reasons for their popularity include simplicity, easy to use, inexpensive hardware implementation, and strong nonlinearity and multimodal behaviors that can be formalized, in many cases, only by experimentally measured data. In a previous paper, we showed that the two-dimensional (2D) LUT controllers and one special type of two-input Mamdani fuzzy controllers are connected in that they have the identical input-output mathematical relation. We also demonstrated how to represent the LUT controllers by the fuzzy controllers. Finally, we showed how to determine the local stability of the LUT control systems. In the present work, we extend these results to the n-dimensional LUT controllers and the special type of the n-input Mamdani fuzzy controllers.

Keywords Fuzzy control · Fuzzy systems · Look-up tables · Stability theory

1 Introduction

The first wave of fuzzy system applications started in the mid 70s with the work of Mamdani and his associates [1, 2] who demonstrated that a family of fuzzy rules could result in a control algorithm that had performance comparable to the conventional industrial controllers. Almost all of the fuzzy system applications at that time followed the mainstream fuzzy control approach-rule-based controllers with fuzzy predicates and reasoning mechanism [3, 4], realizing nonlinear PI, PD or

D. Filev (✉)
Research & Innovation Center, Ford Motor Company, Dearborn, MI, USA
e-mail: dfilev@ford.com

H. Ying
Department of Electrical and Computer Engineering, Wayne State University, Detroit, MI, USA
e-mail: hao.ying@wayne.edu

© Springer International Publishing Switzerland 2017
J. Kacprzyk et al. (eds.), *Granular, Soft and Fuzzy Approaches for Intelligent Systems*, Studies in Fuzziness and Soft Computing 344,
DOI 10.1007/978-3-319-40314-4_10

PID-like control strategies. Most of these works were focused on solving specific control problems, e.g. climate control (Matsushita), subway control (Hitachi), dishwasher and locomotive wheel slip control (General Electric), control of prepaint anticorrosion process (Ford Motor Company), vehicle transmissions (Honda & Nissan), etc. [5–9], Utilizing some of the main advantages of fuzzy control— implementation of intuitive control strategies based on human experience, no requirements for an explicit plant model, rapid prototyping of the control algorithm [10]—these applications gained quick success.

Today, the synergy between fuzzy control, neural networks, evolutionary computing, machine learning, probabilistic/possibilistic reasoning, bio-inspired computational intelligence methodologies, and other soft-computing theory establishes the foundations of a broader control area—intelligent control. Fuzzy control methods are also widely used in conjunction with the conventional ("hard computing") control, diagnosis, pattern recognition, signal processing, knowledge based algorithms and systems where they are introduced within the framework of heuristic strategies at a higher control level (supervisory control, formalization of heuristic task and goals) or/and synergistically with control algorithms that require subjective information, which can be difficult to formalize within the framework of conventional controllers [11].

Various techniques for designing and tuning fuzzy control algorithms have been developed to improve the robustness and tuning of the parameters and the structure of the fuzzy control rule-base by using the similarity between the fuzzy control, PID control, and sliding mode control [12–14]. Significant efforts have been made to rigorously derive and study analytical structure of fuzzy controllers, i.e. the mathematical relationship between the input and output of a fuzzy controller. Precise understanding of the structure is fundamentally important because it can enable one to analyze and design fuzzy control systems more effectively with the aid of conventional control theory [14–18]. The analytical structure is determined by a fuzzy controller's components including input fuzzy sets, output fuzzy sets, fuzzy rules, fuzzy inference, fuzzy logic operators, and defuzzifier. Different component choices obviously result in different analytical structures [2, 11, 16].

In the past 15 years or so, research on improving the performance of fuzzy control algorithms, stability analysis, and systematic design of fuzzy controllers has focused on Takagi-Sugeno method [19] and input-output models of the plant. The Takagi-Sugeno model is a generalization of the gain-scheduling concept—instead of linearizing at a single operating point it enables linearization in multiple vaguely defined regions of the state space [20]. Owing to the fuzzy decomposition the nonlinear system is represented by a polytopic nonlinear structure of coupled linear models that has the property of a universal approximator. The polytopic representation establishes sufficient stability conditions for the TS system using a common Lyapunov function for a set of Lyapunov inequalities [21]. The problem of stability and synthesis is transformed to a convex program. A systematic design methodology that is based on solving LMIs (linear matrix inequalities) has been developed [22]. The TS approach with its strong theoretical foundations was able to overcome the major critics regarding the lack of analyticity of the fuzzy control and made possible to address all major topics of modern control theory. However, despite of the progress towards development of formal analytical model-based approaches for designing fuzzy control systems, most of the practical applications remained centered around heuristic rule-base control. It seems that this observation only confirms the original assertion of Mamdani who introduced fuzzy logic control in 1974 as a powerful tool to "convert heuristic control rules stated by a human operator into an automatic control strategy" [1].

In this chapter we are focusing on a new direction of application of Mamdani controllers that, we believe, has received little, if any, attention. We analytically explore the relationship between one special type of Mamdani fuzzy controllers and the multi-dimensional look-up table (LUT) controllers—one of the most widely used practical engineering tools in industry, especially in automotive engineering—and derive conclusions that contribute to the analysis of the look-up tables based control systems. Our approach is inspired by the similarity between the fuzzy controllers and look-up tables and uses the theory of fuzzy controllers to bring new light to the look-up based engineering technique that is usually considered a low tech and "black art" type control tool. In some sense our approach is just the opposite to the mainstream fuzzy control literature which uses the conventional control theory instrumentarium to explain, analyze, and further develop fuzzy control. Our approach benefits from the great body works on fuzzy control to derive new knowledge and to provide a new interpretation of the look-up type controllers.

LUTs are used in engineering applications as arrays of data that describe relationships between variables. In a broad sense they represent "pseudo-equations to make up for a lack of 'real' equations or perhaps to replace complicated equations with simpler ones" [23, 24]. For example, the vehicle control systems employee thousands of LUTs that contain calibration parameters or define control actions under different operating conditions. The most popular LUTs are the two-dimensional (2D) tables that define the values of one dependent (output) variable for different combinations of two independent (input) variables and even more single dimensional LUTs. They are used as feedforward controllers or as containers for calibrating or gain-scheduling parameters for feedback controllers. The reasons for the popularity of the LUTs in the automotive industry are the strong nonlinearity and multimodal behaviors in the powertrain that can be (in many cases) formalized only by experimentally measured data under different operating conditions. In addition, the LUTs are computationally effective, and can be easily interpreted, visualized, and tuned. Two typical LUTs representing the fuel injection time and the ignition advance at different values of the manifold absolute pressure (MAP) and the engine speed are as follows:

The LUT output is obtained by interpolation (usually linear) in both directions, resulting in a bilinear mapping that will be discussed in more details below. In the following we will show that the output of the LUTs is identical to the output of a special type of Mamdani FLCs.

The chapter is organized in two parts. The first part analyzes the relationship between the Mamdani controllers and the LUT controllers. The main result is a theorem proving the equivalence between one special type of Mamdani controllers with m input variables and the m-dimensional LUT. This result provides the framework for describing and analyzing the LUT as fuzzy controllers. In the second part we show how to determine the local stability of feedback control systems involving the m-dimensional LUT controllers. These findings extend our previous results on the 2D cases [25].

2 On the Relationship Between the Mamdani FLCs and the LUT Controllers

2.1 A Special Type of Mamdani FLCs

The general class of Mamdani FLCs is a rule-base type system that has m input variables, designated as $x_i(n)$, $i = 1, 2, ..., $ m, where n signifies sampling instance. $x_i(n)$ may be a state variable or an input variable computed using the current and/or historical output of a dynamic plant to be controlled (e.g., $y(n)$ and $y(n-1)$) as well as target output signal $S(n)$. This means the input space to be m-dimensional. $x_i(n)$ is multiplied by a scaling factor k_i, resulting in the scaled input variable. For simplicity, we will use $x_i(n)$ to represent the scaled variable. This will not cause confusion because only the scaled input variables will be needed in the rest of the chapter. The universe of discourse for $x_i(n)$ is partitioned into M_i intervals. Like most FLCs in the literature, each interval has at least one fuzzy set defined over it. The j-th fuzzy set of $x_i(n)$ is designated as $\tilde{A}_{i,j}$ whose membership function is denoted $\mu_{\tilde{A}_{i,j}}(x_i)$. $\tilde{A}_{i,j}$ can be any types. The fuzzy controller uses a total of

$$M = \prod_{i=1}^{m} M_i$$ fuzzy rules, each of which is in the following format:

$$\text{IF } x_1(n) \text{ is } \tilde{A}_{1,I_1} \text{ AND } ... \text{ AND } x_m(n) \text{ is } \tilde{A}_{m,I_m} \text{ THEN } u(n) \text{ is } \tilde{V}_k \qquad (1)$$

where the output fuzzy sets $\tilde{V}_k$, k = 1, ..., M, cover the universe of $u(n)$. The membership functions of $\tilde{V}_k$ are denoted $\mu_{\tilde{V}_k}(u)$ and are limited to the singleton type. That is, $\tilde{V}_k$ is nonzero only at one location in the universe of discount for $u(n)$ and the nonzero value is designated as V_k. The fuzzy AND operator is the product operator

$$\tau_h(\mathbf{x}) = \prod_{j=1}^{M} \mu_{\tilde{A}_{j,I_j}}(x_j) \qquad (2)$$

where $\mathbf{x} = [x_1(n) \cdots x_m(n)]$ to define the degrees of firing the rules. As for reasoning, any fuzzy inference method may be used in the rules. It will produce the same inference outcome because the output fuzzy sets are of the fuzzy singleton type (we'll limit the discussion to the case of fuzzy singleton; the extension to fuzzy sets of general shape can be found in [14]). The popular centroid defuzzifier is employed to combine the inference outcomes of the individual rules:

$$u(n) = \frac{\sum\limits_{h=1}^{M} \tau_h(\mathbf{x}) \cdot V_h}{\sum\limits_{h=1}^{M} \tau_h(\mathbf{x})} \tag{3}$$

Here, $\tau_h(x)$ is the resulting membership of executing all the fuzzy logic AND operations in the h-th rule whereas V_h signifies the nonzero value of the singleton output fuzzy set in the rule.

The membership functions of the input fuzzy sets can be of general shape. The only constraint on their selection is to guarantee a complete coverage of the Cartesian product space of all the input variables. Expression (3) defines a deterministic mapping between the inputs and the output of the Mamdani FLCs. For finite universes of the inputs and output, which is always the case for real-world applications, the mapping can be approximated by a LUT. For a predefined rule base and membership functions the LUT can be calculated in advance as part of the FLC design process. The output of the FLC can be inferred from the LUT by interpolation. This simple LUT realization can be applied to any type of controller but it is especially effective in the case of Mamdani FLC because it eliminates the tedious calculations of the degrees of firing using (2)—an operation that might require significant computational resource and time. Almost all references regarding the LUT in the literature on fuzzy control followed the pattern described above— the LUTs were considered as implementation tools approximating the FLCs and their properties have not been analyzed or discussed in the framework of fuzzy control.

In the following we'll show that under certain assumptions the LUTs can be identical to the FLCs, and that implementation of specific FLCs can be computationally effective and simple, comparable to the implementation of the PID controllers. We'll also show that the equivalence between the FLC and LUT can be used to introduce a systematic approach to the local stability analysis of LUT controllers.

In order to simplify the notations we'll first limit the discussion to the 2D case, i.e., assuming two input variables $x_1(n)$ and $x_2(n)$ with the corresponding universes partitioned into intervals covered by fuzzy sets $\widetilde{A}_{1,1}, \ldots, \widetilde{A}_{1,n_1}$ and $\widetilde{A}_{2,1}, \ldots, \widetilde{A}_{1,n_2}$ since this type of FLCs covers the most common cases of PI- and PD-like FLCs. Results will be further generalized to multiple input variables.

The maximal number of rules that are determined by this partitioning is $n = n_1 \times n_2$:

$$\text{IF } x_1(n) \text{ is } \tilde{A}_{1,1} \text{ AND } x_2(n) \text{ is } \tilde{A}_{2,1} \text{ THEN } u(n) \text{ is } V_{11}$$

$$\text{IF } x_1(n) \text{ is } \tilde{A}_{1,1} \text{ AND } x_2(n) \text{ is } \tilde{A}_{2,2} \text{ THEN } u(n) \text{ is } V_{12}$$

$$\cdots \tag{4}$$

$$\text{IF } x_1(n) \text{ is } \tilde{A}_{1,n_1} \text{ AND } x_2(n) \text{ is } \tilde{A}_{2,n_2} \text{ THEN } u(n) \text{ is } V_{n_1 n_2}$$

In this work we'll make one additional assumption on the type of the fuzzy sets $\tilde{A}_{1,1}, \ldots, \tilde{A}_{1,n_1}$ and $\tilde{A}_{2,1}, \ldots, \tilde{A}_{1,n_2}$—they are defined by the normal triangular membership functions. "Normal" means for any values of $x_1(n)$ and $x_2(n)$ the corresponding membership values of the two neighboring fuzzy sets sum to one (Fig. 1 and Table 1). That is

$$\mu_{\tilde{A}_{i,j}}(x_i) + \mu_{\tilde{A}_{i,j+1}}(x_i) = 1, \quad j = \{1, \ldots, n_i - 1\}, i = \{1, 2\}$$

These membership functions are analogous to the concept of B-splines [26]. One can see from Fig. 1 that the normality assumption implies that for any input value $x_1(n)$(or $x_2(n)$) at least one but no more than two of the corresponding membership grades $\tilde{A}_{1,s}$ and $\tilde{A}_{1,s+1}$ (or $\tilde{A}_{2,t}$ and $\tilde{A}_{2,t+1}$) are nonzero and consequently at least one but no more than four of the fuzzy rules (4), including $\tilde{A}_{1,s}$, $\tilde{A}_{1,s+1}$, $\tilde{A}_{2,t}$, and $\tilde{A}_{2,t+1}$, will have nonzero degrees of firing. The normality assumption also means that the membership functions of the input variables are uniquely defined by the

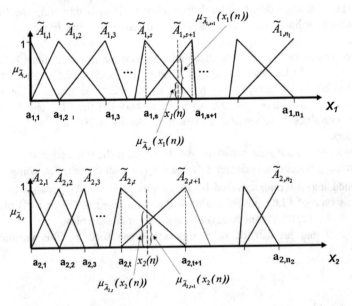

Fig. 1 Normalized membership function of the fuzzy sets of input variables $x_1(n)$ and $x_2(n)$ that satisfy the normality requirement

Table 1 Sample LUTs representing fuel injection time and ignition advance for different engine operating conditions—manifold absolute pressure (MAP) and engine speed [31]

	Speed (RPM)		
MAP (kPA)	500	1000	1500
50	4.02	4.06	4.09
40	3.36	3.39	3.36
30	3.24	3.27	3.31
20	3.04	2.89	3.04
	Speed (RPM)		
MAP (kPA)	500	1000	1500
50	10	16	18
40	9.4	15.7	17.3
30	8.6	15	16.8
20	7.9	14.5	15.9

universe parameters $a_{1,s}$ and $a_{2,t}$, $s = \{1, \ldots, n_1 - 1\}$, $t = \{1, \ldots, n_2 - 1\}$, matching their maxima.

2.2 Relationship Between the Mamdani FLCs and the LUT Controllers

Because of the popularity of 2D LUT controllers in real-world applications, we will first show that if a 2D LUT that has as entries the singleton values $V_{s,t}$, and with rows and columns defined by the universe parameters $a_{1,s}$ and $a_{2,t}$ (Table 2) then the LUT output will be identical to the above fuzzy controllers. In other words, a two-input Mamdani FLC with the normal triangular membership functions of the inputs, singleton consequents and the product AND operator is identical to a 2D LUT that has rows, columns and entries that are defined by the corresponding parameters of the membership functions and the singleton consequents of the FLC.

Table 2 LUT with rows and columns defined by the universe parameters $a_{1,s}$, $a_{2,t}$, and entries— the singleton consequents $V_{s,t}$, $s = \{1, \ldots, n_1 - 1\}$, $t = \{1, \ldots, n_2 - 1\}$

x_1									
x_2	a_{2,n_2}								V_{n_1,n_2}
	a_{2,n_2-1}								
	$\ldots$								
	$a_{2,\,t+1}$				$V_{s,\,t+1}$	$V_{s+1,\,t+1}$			
	$a_{2,\,t}$				$V_{s,\,t}$	$V_{s+1,\,t}$			
	$\ldots$								
	$a_{2,2}$								
	$a_{2,1}$	$V_{1,1}$	$V_{1,2}$						
		$a_{1,1}$	$a_{1,2}$	$\ldots$	$a_{1,\,s}$	$a_{1,\,s+1}$	$\ldots$	a_{1,n_1-1}	a_{1,n_1}

According to Fig. 2 for a set of arbitrarily chosen $x_1(n)$ and $x_2(n)$ we get for the degrees of firing of the affected rules:

$$\tau_{s,t}(\mathbf{x}) = \mu_{\tilde{A}_{1,s}}(x_1) \cdot \mu_{\tilde{A}_{2,t}}(x_2) = \frac{a_{1,s+1} - x_1(n)}{a_{1,s+1} - a_{1,s}} \times \frac{a_{2,t+1} - x_2(n)}{a_{2,t+1} - a_{2,t}}$$

$$\tau_{s,t+1}(\mathbf{x}) = \mu_{\tilde{A}_{1,s}}(x_1) \cdot \mu_{\tilde{A}_{2,t+1}}(x_2) = \frac{a_{1,s+1} - x_1(n)}{a_{1,s+1} - a_{1,s}} \times \frac{x_2(n) - a_{2,t}}{a_{2,t+1} - a_{2,t}}$$

$$\tau_{s+1,t}(\mathbf{x}) = \mu_{\tilde{A}_{1,s+1}}(x_1) \cdot \mu_{\tilde{A}_{2,t}}(x_2) = \frac{x_1(n) - a_{1,s}}{a_{1,s+1} - a_{1,s}} \times \frac{a_{2,t+1} - x_2(n)}{a_{2,t+1} - a_{2,t}} \qquad (5)$$

$$\tau_{s+1,t+1}(\mathbf{x}) = \mu_{\tilde{A}_{1,s+1}}(x_1) \cdot \mu_{\tilde{A}_{2,t+1}}(x_2) = \frac{x_1(n) - a_{1,s}}{a_{1,s+1} - a_{1,s}} \times \frac{x_2(n) - a_{2,t}}{a_{2,t+1} - a_{2,t}}$$

The four firing levels are positive and sum to one. Therefore, the defuzzifier (3) makes the output of the FLC $u(n)$ to interpolate between the corresponding singletons according to the current values of the degrees of firing as functions of the current input values $x_1(n)$ and $x_2(n)$:

$$u(n) = \tau_{s,t}(\mathbf{x})V_{s,t} + \tau_{s,t+1}(\mathbf{x})V_{s,t+1} + \tau_{s+1,t}(\mathbf{x})V_{s+1,t} + \tau_{s+1,t+1}(\mathbf{x})V_{s+1,t+1} \quad (6)$$

Alternatively, from the grid representation of the LUT in Table 1 (Fig. 2), we can obtain $u(n)$ by interpolating first between $V_{s,t}$ and $V_{s+1,t}$, and between $V_{s,t+1}$ and $V_{s+1,t+1}$ along $x_1(n)$ axis. The intermediate interpolated values v_t and v_{t+1}(the order of interpolation does not matter) are

$$v_t = \frac{a_{1,s+1} - x_1(n)}{a_{1,s+1} - a_{1,s}} V_{s,t} + \frac{x_1(n) - a_{1,s}}{a_{1,s+1} - a_{1,s}} V_{s+1,t} \qquad (7)$$

Fig. 2 Grid representation of the LUT of Table 2

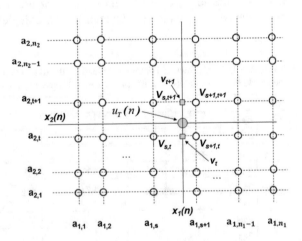

$$v_{t+1} = \frac{a_{1,s+1} - x_1(n)}{a_{1,s+1} - a_{1,s}} V_{s,t+1} + \frac{x_1(n) - a_{1,s}}{a_{1,s+1} - a_{1,s}} V_{s+1,t+1} \tag{8}$$

Similarly, by interpolating between v_t and v_{t+1} along $x_2(n)$ axis and substituting for v_t and v_{t+1} according to (8) and (9) we get for the interpolated value $u_T(n)$ that is inferred by the LUT:

$$
\begin{aligned}
u_T(n) =\ & \frac{a_{2,t+1} - x_2(n)}{a_{2,t+1} - a_{2,t}} v_t + \frac{x_2(n) - a_{2,t}}{a_{2,t+1} - a_{2,t}} v_{t+1} \\
=\ & \frac{a_{2,t+1} - x_2(n)}{a_{2,t+1} - a_{2,t}} \left(\frac{a_{1,s+1} - x_1(n)}{a_{1,s+1} - a_{1,s}} V_{s,t} + \frac{x_1(n) - a_{1,s}}{a_{1,s+1} - a_{1,s}} V_{s+1,t} \right) \\
& + \frac{x_2(n) - a_{2,t}}{a_{2,t+1} - a_{2,t}} \left(\frac{a_{1,s+1} - x_1(n)}{a_{1,s+1} - a_{1,s}} V_{s,t+1} + \frac{x_1(n) - a_{1,s}}{a_{1,s+1} - a_{1,s}} V_{s+1,t+1} \right) \\
=\ & \frac{a_{2,t+1} - x_2(n)}{a_{2,t+1} - a_{2,t}} \frac{a_{1,s+1} - x_1(n)}{a_{1,s+1} - a_{1,s}} V_{s,t} + \frac{a_{2,t+1} - x_2(n)}{a_{2,t+1} - a_{2,t}} \\
& \frac{x_1(n) - a_{1,s}}{a_{1,s+1} - a_{1,s}} V_{s+1,t} + \frac{x_2(n) - a_{2,t}}{a_{2,t+1} - a_{2,t}} \frac{a_{1,s+1} - x_1(n)}{a_{1,s+1} - a_{1,s}} V_{s,t+1} \\
& + \frac{x_2(n) - a_{2,t}}{a_{2,t+1} - a_{2,t}} \frac{x_1(n) - a_{1,s}}{a_{1,s+1} - a_{1,s}} V_{s+1,t+1}.
\end{aligned}
\tag{9}
$$

By comparing with expression (6) we finally obtain:

$$u_T(n) = \tau_{s,t}(\mathbf{x}) V_{s,t} + \tau_{s,t+1}(\mathbf{x}) V_{s,t+1} + \tau_{s+1,t}(\mathbf{x}) V_{s+1,t} + \tau_{s+1,t+1}(\mathbf{x}) V_{s+1,t+1} = u(n)$$

Therefore, the FLC and the LUT produce the same results.

The equivalence between the FLC and the LUT suggests a comprehensive representation and effective computational realization of the FLC with the normal triangular membership functions. We'll illustrate the opportunity for simplifying the implementation of the FLC based on its equivalence to the LUT on the showcase of a simple PI/PD-like FLC. This FLC uses the rule base (5) and has two inputs—the scaled error between the set point and the plant output, $E(n) = K_e e(n)$ and the scaled difference of the error, $\Delta E(n) = K_{\Delta e} \Delta e(n) = K_{\Delta e}(e(n) - e(n-1))$, and one output that coincides with the control variable $u(n)$ (i.e., PD-like FLC) or its difference $\Delta u(n) = u(n) - u(n-1)$ (i.e., PI-like FLC) [14]. A prototypical set of rules of this FLC can be derived from the meta-rules defining a common sense control strategy:

• If $E(n)$ and $\Delta E(n)$ are zero, then maintain present control output
• If $E(n)$ is tending to zero at a satisfactory rate, then maintain present control output
• If $E(n)$ is not self-correcting, then a nonzero $\Delta u(n)$ is added to present control output, depending on the sign and magnitude of $E(n)$ and $\Delta E(n)$

Table 3 Example of a rule base of the Mamdani FLC

		$E(n)$	$\rightarrow$	
		N_e	Z_e	P_e
$\Delta E(n)$	N_d	N^*	NM^*	Z^*
$\downarrow$	Z_d	N^*	Z^*	P^*
	P_d	Z^*	PM^*	P^*

Table 3 shows an example of a PI-like Mamdani FLC rule base that is based on partitioning of the universes of $E(n)$ and $\Delta E(n)$ into three fuzzy sets—Negative (Ne), Zero (Ze), and Positive (Pe) for $E(n)$, and Negative (Nd), Zero (Zd), and Positive (Pd) for $\Delta E(n)$,and fuzzy singletons (real values) Negative ($N^* = -1$), Negative Medium ($NM^* = -0.4$), Zero ($Z^* = 0$), Positive Medium ($PM^* = 0.4$), and Positive ($P^* = 1$) that are defined on the $\Delta u(n)$ universe:

One can easily transfer the rule base in Table 3 to the generic rule base format (4) by letting

$$\tilde{A}_{1,1} = N_e; \; \tilde{A}_{2,1} = N_d; \; \tilde{V}_{1,1} = N^*$$
$$\tilde{A}_{1,2} = N_e; \; \tilde{A}_{2,2} = Z_d; \; \tilde{V}_{2,2} = N^*$$
$$\tilde{A}_{1,3} = N_e; \; \tilde{A}_{2,3} = P_d; \; \tilde{V}_{3,3} = Z^*$$
$$\tilde{A}_{1,4} = Z_e; \; \tilde{A}_{2,4} = N_d; \; \tilde{V}_{4,4} = N^*$$
$$\tilde{A}_{1,5} = Z_e; \; \tilde{A}_{2,5} = Z_d; \; \tilde{V}_{5,5} = Z^*$$
$$\tilde{A}_{1,6} = Z_e; \; \tilde{A}_{2,6} = P_d; \; \tilde{V}_{6,6} = P^*$$
$$\tilde{A}_{1,7} = P_e; \; \tilde{A}_{2,7} = N_d; \; \tilde{V}_{7,7} = Z^*$$
$$\tilde{A}_{1,8} = P_e; \; \tilde{A}_{2,8} = Z_d; \; \tilde{V}_{8,8} = P^*$$
$$\tilde{A}_{1,9} = P_e; \; \tilde{A}_{2,9} = P_d; \; \tilde{V}_{9,9} = P^*$$

If we consider the normal triangular membership functions of $E(n)$ and $\Delta E(n)$ in Fig. 3 we can replace the rule base formulation (Table 3) by the LUT in Table 4.

The output of this LUT equivalent form of the FLC can be calculated by only one line of MATLAB code—line 11 in Fig. 5.

The equivalent LUT form of the FLC combines all the advantages of the FLC with the efficiency and transparency of the LUT. It allows for a computational

Fig. 3 Normal triangular membership functions of the fuzzy sets of the error $E(n)$ and its difference $\Delta E(n)$

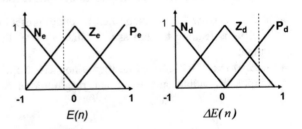

```
% LUT FLC
% FLC Definition
% Universes of Error e and Error Difference de
E=[-1 0 1];
DE=[-2 0 2];
% Control Singletons (each row corresponds to one value of de)
U=[-1 -.4 0; -1 0 1; 0 0.4 1];
% End of FLC Definition
% FLC Output uk for given values of the erorr ek and error difference dek
uk=interp2(E,DE,U,ek,dek);
```

Fig. 4 Sample MATLAB implementation of the LUT equivalence of the FLC. The actual FLC is calculated in last line

Table 4 The LUT that is equivalent to the Mamdani FLC with the rule base by Table 3 and the membership functions by Fig. 4

		$E(n)$	$\rightarrow$	
		-1	**0**	**1**
$\Delta E(n)$	**−2**	-1	−.0.4	0
$\downarrow$	**0**	-1	0	1
	2	0	0.4	1

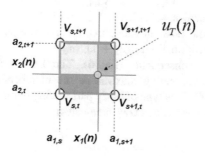

Fig. 5 Four sub-rectangles partitioning the rectangle $(a_{1,s+1} - a_{1,s}) \times (a_{2,t+1} - a_{2,t})$

efficient implementation of the FLC that does not require dealing with the rule base, rules firing, and defuzzification, while exactly preserving all the properties of the FLC.

From Fig. 2 and expression (9) we can see that the interpolated value of the 2D LUT that is inferred for two arbitrary inputs $x_1(n)$ and $x_2(n)$ is essentially the weighted average of the nodes $V_{s,t}$, $V_{s+1,t}$, $V_{s,t+1}$, and $V_{s+1,t+1}$ over the normalized areas $\frac{a_{1,s+1} - x_1(n)}{a_{1,s+1} - a_{1,s}} \cdot \frac{a_{2,t+1} - x_2(n)}{a_{2,t+1} - a_{2,t}}$, $\frac{a_{1,s+1} - x_1(n)}{a_{1,s+1} - a_{1,s}} \cdot \frac{x_2(n) - a_{2,t}}{a_{2,t+1} - a_{2,t}}$, $\frac{x_1(n) - a_{1,s}}{a_{1,s+1} - a_{1,s}} \cdot \frac{a_{2,t+1} - x_2(n)}{a_{2,t+1} - a_{2,t}}$, and $\frac{x_1(n) - a_{1,s}}{a_{1,s+1} - a_{1,s}} \cdot \frac{x_2(n) - a_{2,t}}{a_{2,t+1} - a_{2,t}}$ of the four sub-rectangles. These sub-rectangles are determined by the partitioning of the rectangle $(a_{1,s+1} - a_{1,s}) \times (a_{2,t+1} - a_{2,t})$ by $x_1(n)$ and $x_2(n)$, and each of them is associated with one of four nodes $V_{s,t}$, $V_{s+1,t}$, $V_{s,t+1}$, and $V_{s+1,t+1}$—see Fig. 6 that illustrates this partitioning. The areas of the rectangles are used as a measure of closeness between the input $(x_1(n), x_2(n))$ and the nodes in the 2D space. It is easy to see that the weights are positive and sum to one.

In the above discussion we demonstrated the equivalence between the FLCs and the LUT considering the commonly used two-inputs, single-output model of a FLC and the 2D LUT that is widely accepted in industry. We now extend this result to cover m-dimensional LUT controllers, which is important both theoretically and practically. That is, we'll use the above observation to derive the expression for the output inferred by an mD LUT and to prove its equivalence to an mD FLC.

Theorem 1 *A Mamdani FLC with m inputs partitioned into (ni + 1), i = {1, 2, ..., m}, fuzzy subsets with the normal triangular membership functions, product AND aggregation of the input fuzzy subsets, and singleton consequents is equivalent to an m-dimensional LUT with grid points defined by the arguments of the maxima of the membership functions of the input variables and grid point entries corresponding to the consequents.*

Proof Assume an mD LUT where the inputs are divided into an m dimensional grid of n_1, n_2, ..., n_m of m dimensional cells defined by the ordered grid points $a_{i,1}, a_{i,2}, \ldots, a_{i,s_i}, a_{i,n_i+1}, i = \{1, 2, \ldots, m\}$ and corresponding functional values (LUT entries) $V_{1,1,\ldots,1}$, $V_{1,1,\ldots,2}, \ldots, V_{1,1,\ldots,s_i}$, $V_{1,1,\ldots,s_i+1}$, $\ldots, V_{1,1,\ldots,n_m+1},\ldots,$ $V_{n_1+1,n_2+1,\ldots,n_m+1}$ where $s_i = \{1, 2, \ldots, n_i\}, i = \{1, 2, \ldots, m\}$ are arbitrary intermediate points. Assume now an mD FLC, i.e. an m-input, single output fuzzy system with inputs partitioned into $(n_1 + 1)$, $(n_2 + 1)$, ..., $(n_m + 1)$ normal fuzzy subsets and singleton consequents (Fig. 6). The LUT grid points and entries coincide with the corresponding maxima of the membership functions and the consequents of the FLC. This type of fuzzy partitioning defines a family of $(n_1 + 1)$ $(n_2 + 1) \ldots (n_m + 1)$ rules with fuzzy predicates:

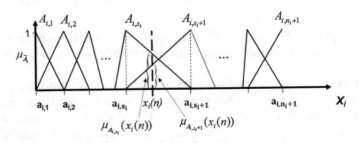

Fig. 6 Membership functions of the fuzzy sets of the i-th FLC input x_i

IF x_1 *is* $A_{1,1}$ *AND* x_2 *is* $A_{2,1}$ *AND* . . . x_m *is* $A_{m,1}$ *THEN* u *is* $V_{1,1,...,1}$

IF x_1 *is* $A_{1,1}$ *AND* x_2 *is* $A_{2,1}$ *AND* . . . x_m *is* $A_{m,2}$ *THEN* u *is* $V_{1,1,...,2}$

. . .

IF x_1 *is* $A_{1,1}$ *AND* x_2 *is* $A_{2,1}$ *AND* . . . x_m *is* A_{m,s_m} *THEN* u *is* $V_{1,1,...,s_m}$

IF x_1 *is* $A_{1,1}$ *AND* x_2 *is* $A_{2,1}$ *AND* . . . x_m *is* A_{m,s_m+1} *THEN* u *is* $V_{1,1,...,s_m+1}$

. . .

IF x_1 *is* $A_{1,1}$ *AND* x_2 *is* $A_{2,1}$ *AND* . . . x_m *is* A_{m,n_m+1} *THEN* u *is* $V_{1,1,...,n_m+1}$

. . .

IF x_1 *is* A_{1,n_1+1} *AND* x_2 *is* A_{2,n_2+1} *AND* . . . x_m *is* A_{m,n_m+1} *THEN* u *is* $V_{n_1+1,n_2+1,...,n_m+1}$

Consider the m dimensional LUT cell C_s formed by $a_{1,s_1} \leq x_1(n) \leq a_{1,s_1+1}$, $a_{2,s_2} \leq x_2(n) \leq a_{2,s_2+1}, a_{m,s_m} \leq x_m(n) \leq a_{m,s_m+1}$, and $s_i = \{1, 2, \ldots, n_i\}, i = \{1, 2, \ldots, m\}$. The cell is defined by the hyperrectangle

$$C_s \equiv (a_{1,s_1+1} - a_{1,s_1}) \times (a_{2,s_2+1} - a_{2,s_2}) \times \ldots \times (a_{m,s_m+1} - a_{m,s_m}).$$

The hyperrectangle C_s has 2^m nodes, $V_{s_1,s_2,...,s_m}, V_{s_1,s_2,...,s_m+1}, \cdots,$ $V_{s_1+1,s_2+1,...,s_m+1}$, which are associated with the LUT entries. The hyperrectangle C_s is divided by the inputs $x_1(n), x_2(n), \ldots,$ and $x_m(n)$ into 2^m sub-hyperrectangles:

$$C_{s_1+1,s_2+1,...,s_m+1} \equiv (a_{1,s_1+1} - x_1(n)) \times (a_{2,s_2+1} - x_2(n)) \times \ldots \times (a_{m,s_m+1} - x_m(n))$$
$$C_{s_1+1,s_2+1,...,s_m} \equiv (a_{1,s_1+1} - x_1(n)) \times (a_{2,s_2+1} - x_2(n)) \times \ldots \times (x_m(n) - a_{m,s_m})$$
$$C_{s_1,s_2,...,s_m} \equiv (x_1(n) - a_{1,s_1}) \times (x_2(n) - a_{2,s_2}) \times \ldots \times (x_m(n) - a_{m,s_m})$$

Each of the sub-hyperrectangles includes one of the nodes of the cell C_s. Using the definition of a volume of a hyperrectangle as a product of its sides [27] (note that this definition of the volume of a hyperrectangle is consistent with the definitions of the area of a rectangle in 2D and the volume of a cuboid in the 3D space) we can express the volume of the cell C_s as a sum of the volumes of the sub-hyperrectangles $C_{s_1,s_2,...,s_m}, C_{s_1,s_2,...,s_m+1}, \cdots, C_{s_1+1,s_2+1,...,s_m+1}$:

$$(a_{1,s_1+1} - a_{1,s_1})(a_{2,s_2+1} - a_{2,s_2}) \ldots (a_{m,s_m+1} - a_{m,s_m}) = (a_{1,s_1+1} - x_1(n))$$
$$(a_{2,s_2+1} - x_2(n)) \ldots (a_{m,s_m+1} - x_m(n)) +$$
$$(a_{1,s_1+1} - x_1(n))(a_{2,s_2+1} - x_2(n)) \ldots (x_m(n) - a_{m,s_m}) + \ldots$$
$$+ (x_1(n) - a_{1,s_1})(x_2(n) - a_{2,s_2}) \ldots (x_m(n) - a_{m,s_m})$$

Similarly to the 2D case, the volumes of the sub-hyperrectangulars reflect the closeness of the input $(x_1(n), x_2(n), \ldots x_m(n))$ to the nodes of C_s. By associating the normalized volumes of the sub-hyperrectangles with the corresponding nodes of

$C_s - V_{s_1, s_2, \ldots, s_m}$ $V_{s_1, s_2, \ldots, s_m + 1}, \ldots, V_{s_1 + 1, s_2 + 1, \ldots, s_m + 1}$—we get the interpolated value that is inferred by the rD LUT:

$$
\begin{aligned}
u_T(n) = \frac{1}{K} \Big(&(a_{1, s_1 + 1} - x_1(n))(a_{2, s_2 + 1} - x_2(n)) \ldots (a_{r, s_r + 1} - x_r(n)) \\
&V_{s_1 + 1, s_2 + 1, \ldots, s_m + 1} + (a_{1, s_1 + 1} - x_1(n))(a_{2, s_2 + 1} - x_2(n)) \\
&(x_r(n) - a_{r, s_r})(V_{s_1 + 1, s_2 + 1, \ldots, s_m}) + \ldots + (x_1(n) - a_{1, s_1}) \\
&(x_2(n) - a_{2, s_2}) \ldots (x_r(n) - a_{r, s_r}) V_{s_1, s_2, \ldots, s_m} \Big)
\end{aligned}
\tag{10}
$$

where the normalizing factor $K = (a_{1, s_1 + 1} - a_{1, s_1})(a_{2, s_2 + 1} - a_{2, s_2}) \ldots (a_{m, s_m + 1} - a_{m, s_m})$ is the volume of the hyperrectangle $(a_{1, s_1 + 1} - a_{1, s_1}) \times (a_{2, s_2 + 1} - a_{2, s_2}) \times \ldots \times (a_{m, s_m + 1} - a_{m, s_m})$. It is easy to see that by combining pair of the terms with all but one elements identical, e.g. the first two terms in (), the weights will sum to one.

Consider now that the same arbitrary input values $x_1(n), x_2(n)$, and $x_m(n)$ where

$$
a_{1, s_1} \le x_1(n) \le a_{1, s_1 + 1}, a_{2, s2} \le x_2(n) \le a_{2, s_2 + 1}, \ldots, a_{m, s_m} \le x_m(n) \le a_{m, s_m + 1},
$$

and $s_i = \{1, 2, \ldots, m - 1\}$, $i = \{1, 2, \ldots, m\}$ are applied to the FLC. As we can see from Fig. 6 only the following 2^m rules:

IF x_1 is A_{1, s_1} AND x_2 is A_{2, s_2} AND $\ldots x_m$ is A_{m, s_m} THEN u is $V_{s_1, s_2, \ldots, s_m}$

IF x_1 is A_{1, s_1} AND x_2 is A_{2, s_2} AND $\ldots x_m$ is $A_{m, s_m + 1}$ THEN u is $V_{s_1, s_2, \ldots, s_m + 1}$

$\cdots$

IF x_1 is $A_{1, s_1 + 1}$ AND x_2 is $A_{2, s_2 + 1}$ AND $\ldots x_m$ is $A_{m, s_m + 1}$ THEN u is $V_{s_1 + 1, s_2 + 1, \ldots, s_m + 1}$

will fire because their antecedents have nonzero membership values.

Following Fig. 6 and assuming a product AND operator the degrees of firing of those rules are:

$$
\tau_{s_1, s_2, \ldots, s_m} = \frac{(x_1(n) - a_{1, s_1})(x_2(n) - a_{2, s_2}) \ldots (x_m(n) - a_{m, s_m})}{(a_{1, s_1 + 1} - a_{1, s_1})(a_{2, s_2 + 1} - a_{2, s_2}) \ldots (a_{m, s_m + 1} - a_{m, s_m})}
$$

$$
\tau_{s_1, s_2, \ldots, s_m + 1} = \frac{(x_1(n) - a_{1, s_1})(x_2(n) - a_{2, s_2}) \ldots (x_m(n) - a_{m, s_m + 1})}{(a_{1, s_1 + 1} - a_{1, s_1})(a_{2, s_2 + 1} - a_{2, s_2}) \ldots (a_{m, s_m + 1} - a_{m, s_m})}
$$

$\cdots$

$$
\tau_{s_1 + 1, s_2 + 1, \ldots, s_m + 1} = \frac{(x_1(n) - a_{1, s_1 + 1})(x_2(n) - a_{2, s_2 + 1}) \ldots (x_m(n) - a_{m, s_m + 1})}{(a_{1, s_1 + 1} - a_{1, s_1})(a_{2, s_2 + 1} - a_{2, s_2}) \ldots (a_{m, s_m + 1} - a_{m, s_m})}
$$

By applying the defuzzification law (3) we obtain for u(n) an expression that is identical to that for the output of the LUT (10):

$$u(n) = \frac{(x_1(n) - a_{1,s_1})(x_2(n) - a_{2,s_2}) \dots (x_m(n) - a_{m,s_m})}{(a_{1,s_1+1} - a_{1,s_1})(a_{2,s_2+1} - a_{2,s_2}) \dots (a_{m,s_m+1} - a_{m,s_m})} V_{s_1,s_2,\dots,s_m}$$
$$+ \frac{(x_1(n) - a_{1,s_1})(x_2(n) - a_{2,s_2}) \dots (x_m(n) - a_{m,s_m+1})}{(a_{1,s_1+1} - a_{1,s_1})(a_{2,s_2+1} - a_{2,s_2}) \dots (a_{m,s_m+1} - a_{m,s_m})} V_{s_1,s_2,\dots,s_m+1} + \dots$$
$$+ \frac{(x_1(n) - a_{1,s_1+1})(x_2(n) - a_{2,s_2+1}) \dots (x_m(n) - a_{m,s_m+1})}{(a_{1,s_1+1} - a_{1,s_1})(a_{2,s_2+1} - a_{2,s_2}) \dots (a_{m,s_m+1} - a_{m,s_m})} V_{s_1+1,s_2+1,\dots,s_m+1}$$

We point out that the properties of the FLCs with the normal triangular membership functions have been studied by a number of researchers. Most of their work have been focused on analyzing the similarity between the FLCs and the linear PI and PD controllers, demonstrating that for special values of the consequents the FLCs are identical to the nonlinear PI or PD controllers (e.g., [14, 16, 28]). A more detailed recent studies of the properties of the fuzzy models with the normal membership functions from the perspective of bilinear systems by Sugeno and Taniguchi [29] were partially the inspiration for our work.

In the next section we'll show the input-output relationship of the LUT controllers and how to determine the local stability of the LUT control systems—a topic the has received very little, if any, attention although a great deal of practical control applications are based on such LUT controllers.

3   Analytical Structure and Local Stability of the LUT Control Systems

We will first focus on the analytical structure of the fuzzy controllers with two input variables and will then generalize the result to the m-dimensional LTU controllers. By simplifying (9) we obtain the analytical structure of the 2D LUT controllers, which is also the 2D FLCs, as follows:

$$
\begin{aligned}
u_T(n) = u(n) = {} & \frac{a_{2,t+1} - x_2(n)}{a_{2,t+1} - a_{2,t}} \frac{a_{1,s+1} - x_1(n)}{a_{1,s+1} - a_{1,s}} V_{s,t} + \frac{a_{2,t+1} - x_2(n)}{a_{2,t+1} - a_{2,t}} \\
& \frac{x_1(n) - a_{1,s}}{a_{1,s+1} - a_{1,s}} V_{s+1,t} + \frac{x_2 - a_{2,t}}{a_{2,t+1} - a_{2,t}} \frac{a_{1,s+1} - x_1(n)}{a_{1,s+1} - a_{1,s}} V_{s,t+1} \\
& + \frac{x_2 - a_{2,t}}{a_{2,t+1} - a_{2,t}} \frac{x_1(n) - a_{1,s}}{a_{1,s+1} - a_{1,s}} V_{s+1,t+1} \qquad (11) \\
= {} & \alpha x_1(n) + \beta x_2(n) + \gamma x_1(n) x_2(n) + \delta
\end{aligned}
$$

where α, β, γ, and δ are constants:

$$\alpha = (a_{2,t+1}V_{s+1,t} - a_{2,t+1}V_{s,t} + a_{2,t}V_{s,t+1} - a_{2,t}V_{s+1,t+1})/K$$
$$\beta = (a_{1,s+1}V_{s,t+1} - a_{1,s+1}V_{s,t} + a_{1,s}V_{s+1,t} - a_{1,s}V_{s+1,t+1})/K$$
$$\gamma = (V_{s,t} - V_{s+1,t} - V_{s,t+1} + V_{s+1,t+1})/K$$
$$\delta = (a_{1,s+1}a_{2,t+1}V_{s,t} + a_{1,s}a_{2,t}V_{s+1,t+1} - a_{1,s}a_{2,t+1}V_{s+1,t} - a_{1,s+1}a_{2,t}V_{s,t+1})/K$$
$$K = (a_{2,t+1} - a_{2,t})(a_{1,s+1} - a_{1,s})$$

That is, the LUT controllers are nonlinear controllers with constant offset term (δ).

Likewise, simplifying (10), one can attain the nonlinear analytical structure of the rD LUT controllers, or equivalently that of the rD FLCs:

$$u_T(n) = u(n) = \sum_{d_i \geq 0} \left(\beta_{d_1 \dots d_r} \prod_{i=1}^{r} x_i^{d_i} \right), \quad \sum_{i=1}^{r} d_i \leq r \qquad (12)$$

where the values of the constants $\beta_{d_1 \dots d_r}$, is determined by all the constant parameters in (10). For concise presentation, we omit their complicated general relations with these parameters. For any specific LUT controller, especially those involving only a handful number of input variables which is mostly the case in practice, $\beta_{d_1 \dots d_r}$, can be easily determined. For instance, when there are two input variables, based on (11), it is obvious that $\beta_{10} = \alpha$, $\beta_{01} = \beta$, $\beta_{11} = \gamma$, and $\beta_{00} = \delta$.

We now turn our attention to the local stability determination of the rD LUT controller regulating a nonlinear dynamic system. Without loss of generality, assume that when the system to be controlled is at the equilibrium point of our interest, $x_1(n) = \dots = x_r(n) = 0$. We want to study the condition for the nonlinear rD LUT control system to be stable at least in the area around the equilibrium point. If both the system and the LUT controller are linearizable at the equilibrium point, then the system stability at that point can be decided by applying Lyapunov's linearization method [30] to the linearized LUT controller and the linearized system. Thus, we obtain the following result:

Theorem 2 *Suppose that the rD LUT controller is used to control a nonlinear system that is linearizable at the equilibrium point. The control system is locally stable (or unstable) at the equilibrium point if and only if the linearized system involving the LUT controller linearized at the equilibrium point*

$$u_T(n) = \beta_{10\dots0}x_1(n) + \beta_{010\dots0}x_2(n) + \cdots + \beta_{0\dots0r}x_r(n)$$

is strictly stable (or unstable).

Proof First, notice that because (12) is a multi-variable polynomial, it is always linearizable at an equilibrium point no matter where it is. Now suppose that the equilibrium point is $x_1(n) = \dots = x_r(n) = 0$. Linearization of (12) is achieved if all the cross-product terms of the variables are dropped because $x_i(n) \approx 0$ for all the variables and hence the higher order cross-product terms are much smaller than $x_i(n)$. This results in

$$u_T(n) = \beta_{10\cdots 0} x_1(n) + \beta_{010\cdots 0} x_2(n) + \cdots + \beta_{0\cdots 0r} x_r(n) + \beta_{0\cdots 0}$$

Since the constant term $\beta_{0\cdots 0}$ does not affect the system stability, it is removed from the stability study. The conclusion naturally follows by using Lyapunov's linearization method. The method states that if the nonlinear control system is continuously differentiable at the equilibrium point and the linearized system is strictly stable (or unstable) at the equilibrium point, then the equilibrium point is locally stable (or unstable) for the original nonlinear system. Q.E.D The linearizability test must be met for the system to be controlled as it is the precondition for the theorem. A test failure only means inapplicability of the theorem; it does not imply system instability. Theorem 2 offers some practically important advantages. First, it is a necessary and sufficient condition. Unlike sufficient conditions or necessary conditions, it is not conservative and is the "tightest" possible stability condition. Second, the theorem can be used not only when the system model is available, but also when it is unavailable but is known linearizable at the equilibrium point (most physical systems are linearizable).

4 Conclusion

In this chapter we expanded our previous work on the relationship between the LUT controllers and one special type of fuzzy controller. We showed that the multidimensional LUT's are closely related to the fuzzy models and can be considered and analyzed, in a broad sense, a special class of fuzzy models.

References

1. Mamdani, E., Assilian, S.: An experiment in linguistic synthesis with a fuzzy logic controller. Int. J. Man Mach. Stud. **7**(1), 1–13 (1975)
2. Kickert, W., Mamdani, E.: Analysis of a fuzzy logic controller. Fuzzy Sets Syst. **12**, 29–44 (1978)
3. Reznik, Leon: Fuzzy Controllers Handbook: How to Design Them, How They Work. Newnes, Oxford (1997)
4. Zadeh, L.: Outline of a new approach to the analysis of complex systems and decision processes. IEEE Trans. Syst. Man Cybern. **3**, 28–44 (1973)
5. Sugeno, M. (ed.): "Industrial Applications of Fuzzy Logic Control.". North Holland, Amsterdam (1985)
6. Schwartz, D.G., Klir, G.J.: Fuzzy logic flowers in Japan. IEEE Spectr. **29**, 32–35 (1991)
7. Schwartz, D.G., Klir, G.J., Lewis III, H.W., Ezawa, Y.: Applications of fuzzy sets and approximate reasoning. Proc. IEEE **82**, 482–498 (1994)

8. Bonissone, P.: Fuzzy logic controllers: An industrial reality. In: Zurada, J., Marks, R., Robinson, C. (eds.) Computational Intelligence Imitating Life, pp. 316–327. IEEE Press, New York, NY (1994)
9. Farinwata, S., Filev, D., Langari, R. (eds.): Fuzzy Control: Synthesis and Analysis. John Wiley & Sons Ltd, London (2000)
10. Filev, D., Syed, F.: Applied control systems: blending fuzzy logic with conventional control systems. Int. J. General Syst. **39**, 395–414 (2010)
11. Eciolaza, L., Taniguchi, T., Sugeno, M., Filev, D., Wang, Y.: PB model-based FEL and its application to driving pattern learning. In: 19th IFAC Congress, Cape Town, Aug 2014
12. Palm, R.: Robust-control by fuzzy sliding mode. Automatica **30**(9), 1429–1437 (1994)
13. Palm, R., Driankov, D., Hellendoorn, H.: Model Based Fuzzy Control: Fuzzy Gain Schedulers and Sliding Mode Fuzzy Controllers. Springer-Verlag, New York (1996)
14. Yager, R., Filev, D.: Essentials of Fuzzy Modeling and Control. John Wiley & Sons, New York (1994)
15. Ying, H.: Practical design of nonlinear fuzzy controllers with stability analysis for regulating processes with unknown mathematical models. Automatica **30**, 1185–1195 (1994)
16. Ying, H.: Fuzzy Control and Modeling. IEEE Press, New York (2000)
17. Ying, H., Filev, D.: Analytical structure characterization and stability analysis for a general class of mamdani fuzzy controllers. In: Zadeh, L.A., Abbasov, A.M., Yager, R.R., Shahbazova, S.N., Reformat, M.Z. (eds.) Recent Developments and New Directions in Soft Computing, pp. 233–247. Springer Verlag (2014)
18. Du, X.Y., Zhang, N.Y., Ying, H.: Structure analysis and system design for a class of Mamdani fuzzy controllers. Int. J. Gen Syst **37**, 83–101 (2008)
19. Takagi, T., Sugeno, M.: Fuzzy identification of systems and its applications to modeling and control. IEEE Trans. Syst. Man. Cybern. **15**, 116–132 (1985)
20. Filev, D.: Fuzzy modeling of complex systems. Int. J. Approximate Reasoning **5**, 281–290 (1991)
21. Filev, D.: Gain scheduling based control of a class of TSK systems. In: Farinwata, S., Filev, D., Langari, R. (eds.) Fuzzy Control: Synthesis and Analysis, pp. 321–334. John Wiley & Sons Ltd, London (2000)
22. Tanaka, K., Wang, H.O.: Fuzzy Control Systems Design and Analysis: A Linear Matrix Inequality Approach. John Wiley & Son, New York (2001)
23. Bernstein, D.: From the editor—Theorem and look-up table. IEEE Control Syst. **24**, 6 (2004)
24. Vogt, M., Muller, N., Isermann, R.: On-line adaptation of grid-based look-up tables using a fast linear regression technique. Trans. ASME **126**, 732–739 (2004)
25. Filev, D., Ying, H.: The look-up table controllers and a particular class of mamdani fuzzy controllers are equivalent—implications to real-world applications. In: Proceedings of 2013 IFSA World Congress and NAFIPS Annual Meeting, Edmonton, Canada, 24–28, June 2013
26. de Boor, C.: A Practical Guide to Splines. Springer-Verlag (1978)
27. Ordonez, C., Omiecinski, E., Navathe, S., Ezquerra, N.: A clustering algorithm to discover low and high density hyper-rectangles in subspaces of multidimensional data. Technical Report; GIT-CC-99-20, Georgia Institute of Technology (1999)
28. Mizumoto, M.: Realization of PID controls by fuzzy control methods. In: Proceedings of IEEE International Conference on Fuzzy Systems, San Diego, CA, 8–12 Mar 1992
29. Taniguchi, T., Sugeno, M.: Stabilization of nonlinear systems with piecewise bilinear models derived from fuzzy if-then rules with singletons. In: Proceedings of the WCCI 2010 IEEE World Congress on Computational Intelligence, pp. 2926–2931, Barcelona, Spain, 18–23 Jul 2010
30. Slotine, J.J., Li, W.P.: Applied Nonlinear Control. Prentice Hall (1991)
31. Wong, P.K., Tam, L.M., Li, K., Wong, H.C.: Automotive engine idle speed control optimization using least squares support vector machine and genetic algorithm. Int. J. Intell. Comput. Cybern. **1**, 598–616 (2008)

FuzzyLP: An R Package for Solving Fuzzy Linear Programming Problems

Pablo J. Villacorta, Carlos A. Rabelo,
David A. Pelta and José Luis Verdegay

Abstract An inherent limitation of Linear Programming is the need to know precisely all the conditions concerning the problem being modeled. This is not always possible as there exist uncertainty situations which require a more suitable approach. Fuzzy Linear Programming allows working with imprecise data and constraints, leading to more realistic models. Despite being a consolidated field with more than 30 years of existence, almost no software has been developed for public use that solves fuzzy linear programming problems. Here we present an open-source R package to deal with fuzzy constraints, fuzzy costs and fuzzy coefficients in linear programming. The theoretical foundations for solving each type of problem are introduced first, followed by code examples. The package is accompanied by a user manual and can be freely downloaded, employed and extended by any R user.

1 Introduction

Linear Programming (LP) is one of the main branches of Operational Research. It is composed of optimization models whose objective function and constraints are linear on the decision variables. Due to their simplicity, they have often been used for solving a wide variety of problems in sciences and engineering, enabling important benefits and savings for companies and organizations. Efficient algorithms exist for this problem, such as simplex, created in 1947 [11].

P.J. Villacorta (✉) · C.A. Rabelo · D.A. Pelta · J.L. Verdegay
Department of Computer Science and Artificial Intelligence,
University of Granada, Granada, Spain
e-mail: pjvi@decsai.ugr.es

C.A. Rabelo
e-mail: carabelo@gmail.com

D.A. Pelta
e-mail: dpelta@decsai.ugr.es

J.L. Verdegay
e-mail: verdegay@decsai.ugr.es

© Springer International Publishing Switzerland 2017
J. Kacprzyk et al. (eds.), *Granular, Soft and Fuzzy Approaches
for Intelligent Systems*, Studies in Fuzziness and Soft Computing 344,
DOI 10.1007/978-3-319-40314-4_11

One limitation of LP is the requirement to know precisely all the parameters of the problem. This is sometimes not possible due to risk or uncertainty in some data, which can be handled using fuzzy numbers. Such situations are very common. Fuzzy Linear Programming (FLP) [17], as a particular case of the more broad field of Fuzzy Convex Optimization [30], allows working with imprecision in both the coefficients and the constraints, yielding more realistic models. We can distinguish three cases:

- Problems with fuzzy constraints problems, where some degree of violation of the constraints is allowed.
- Problems with fuzzy costs, where the coefficients of the objective function are fuzzy.
- Problems with fuzzy coefficients, where the coefficients of the constraints are fuzzy.

although it is usual to find problems that combine more than one of these.

Consider the following motivating examples. In [5], the distribution of frozen food along a network, which could be solved with a linear program in case all the parameter were precisely known, is approached as an FLP problem with fuzzy constraints. The reason is that time-related parameters such as the time needed by the refrigerated truck to move between locations, the time the truck doors remain open when unloading, or the time windows imposed by the clients are not precisely known, hence the constraints associated to these uncertain quantities should be treated as soft constraints. In [7], the main aim is to design a diet which enables an animal of certain characteristics to achieve certain level of daily weight gain. Parameters involved in the problem, such as the exact amount of each type of food owned by the farmer, or the amount eaten by each animal are not known precisely by the farmer since the livestock is treated as a whole. The constraints on the maximum quantities of each nutrient each animal should eat are fuzzy for the same reason, and also because some tolerance is permitted if it leads to cheaper diets. Finally, in [23], the problem of supply chain planning is tackled via a fuzzy MILP approach considering that, in a real scenario, quantities such as the costs corresponding to idleness, raw material acquisition, inventory holding, demand backlog and transport cannot be measured easily since they imply human perception for their estimation, hence they are modeled using fuzzy numbers. The same applies to the production time, the production capacity, and some others.

A list of applications of FLP can be found in [26]; we summarize some of them below.

- Agricultural economy: water usage and water scheduling in agriculture, diet problems, optimization of farms structure, allocation of regional resources.
- Banking: assessing financial assets, portfolio problems, investment.
- Environment: regulation of air pollution, energy production models.
- Manufacturing: aggregated production scheduling, machine optimization, optimal allocation for metal manufacturing, optimal system design, raw oil processing.
- Personnel management and coordination.
- Transportation problems, track routing problems.

Further examples include industrial production planning [35], optimal water allocation with environmental constraints [34], optimal cattle diets [7], supply chain management under uncertainty [23], optimization of football team resources to maximize performance [6], and energy management [29], just to cite a few.

Although a very large number of approaches can be found in the literature, software solutions are scarce. As highlighted in [29], researchers usually focus on toy examples to demonstrate their novel mathematical solution methods. In [29], a proposal on a fuzzy version of GAMS [28], a popular algebraic modeling language for crisp mathematical programming, is briefly outlined but no further development was done. Apart from this, there exist (a) a non-general software aimed at solving a concrete problem, such as SACRA [8] for cattle diet optimization, and (b) a decision support system, called PROBO [9], written in Pascal which is difficult to integrate with current software environments.

This contribution is aimed at partially filling this gap by presenting a ready-to-use implementation in a modern language so that existing FLP methods can be used by the R community without much effort.[1] At the same time, we hope our work will contribute to further software developments in this direction to make fuzzy mathematical programming models in general readily available for researchers and practitioners from other areas, spreading this methodology beyond the fuzzy community.

The remainder of the present work is structured as follows. Section 2 provides background on some key concepts concerning fuzzy numbers, although the reader is assumed already familiar with the foundations of fuzzy sets. Section 3 describes the mathematical models implemented in our package, along with fragments of R code showing their corresponding implementation and usage. Section 4 is devoted to conclusions and further work.

A Note on R

The R programming language [25, 37] is oriented to data manipulation, data analysis and graphics. It has its origin in the S and S-Plus languages. Its modern version was established in 1988 and is basically an open-source implementation of the S language. The website of the R project[2] is the main information source and collects extensive documentation, both official and contributed. R is similar to MATLAB® in that it provides many operators for doing computations with vectors and matrices in a seamless way, as well as a large collection of built-in mathematical and statistical functions for computing with data. As happens with other mathematical languages and environments, it can be used either interactively from a console, or by running R scripts which implement a more elaborated workflow.

Possibly the two features which have contributed most to the fast adoption of R are the following. First, it is open-source, hence the users can access the source code of all R functionality including the built-in core functions. Second, it can be extended by using *packages*. A package is a collection of functions, data structures such as classes (as R is object-oriented) and data gathered together to accomplish a concrete

---

[1]FuzzyLP can be downloaded from http://cran.r-project.org/package=FuzzyLP.
[2]http://www.r-project.org.

task. A central repository called CRAN[3] (Comprehensive R Archive Network) stores all packages contributed by R users. When one develops a package along with the necessary documentation, it can be submitted to CRAN after passing some minimal quality and usability standards. From that moment, anyone can easily download and use it in his/her own programs, as well as access the package source code and modify or extend it in any way. As a result, a fast dissemination of novel, highly specialized methods implemented directly by their authors is achieved. In the long run, this practice, known as reproducible research, allows for a faster advancement in many research areas as new approaches are built on top of (and can be easily compared with) existing proposals that are readily available. As of August 2016 there are more than 8700 contributed R packages on CRAN, including FuzzyLP. The packages are aimed at either implementing statistical methods, giving access to data or hardware, or serving as a complement to textbooks.

Two contributed packages have been used in implementing our FuzzyLP package:

- The package FuzzyNumbers [18] provides classes and methods for working with fuzzy numbers. More precisely, it allows representing generic fuzzy numbers as well as some particular types (triangular, trapezoidal, piecewise linear). It offers several defuzzification functions as well as functions to approximate general fuzzy numbers using piecewise linear fuzzy numbers. Basic arithmetic can also be done with fuzzy numbers. Finally, it also has plotting facilities. According to the author, future versions may include random fuzzy number generation, aggregation and sorting.
- Package ROI (R Optimization Infrastructure [33]). Since FLP ultimately relies on crisp LP solving algorithms, this package has been used to accomplish such task. A large number of optimization packages are available in R; see the R Task View on Optimization.[4] Among them, ROI is an attempt to provide a unified interface for any optimization problem. For this reason, along with the possibility that future versions of our FuzzyLP package need to deal with other non-linear fuzzy optimization problems which eventually rely on more complex crisp solvers that ROI already offers, we have chosen this package.

2 Fuzzy Numbers

We assume the reader is familiar with the basics of fuzzy sets and fuzzy numbers, and therefore we only review those concepts that are highly relevant for our later exposition. The interested reader may refer to [15, 22] for an introduction to the topic.

---

[3]http://cran.r-project.org/web/packages/available_packages_by_name.html.

[4]http://cran.r-project.org/web/views/Optimization.html.

Definition 1 A fuzzy number [14] $\tilde{A}$ is a convex and normalized subset of the real line such that

(i) $\forall \alpha \in [0, 1], \tilde{A}_\alpha = \{x \in \mathbb{R} | \mu_{\tilde{A}} \geq \alpha\}$ (the α-cuts of $\tilde{A}$) is a convex set.
(ii) $\mu_{\tilde{A}}$ is upper semi-continuous
(iii) $Supp(\tilde{A}) = \{x \in \mathbb{R} | \mu_{\tilde{A}} > 0\}$ is a bounded set on $\mathbb{R}$.

Definition 2 Given $r < u \leq U < R \in \mathbb{R}$, a Trapezoidal Fuzzy Number (TrFN) $\tilde{u} = (r, u, U, R)$ is defined as:

$$\mu_{\tilde{u}}(x) = \begin{cases} \frac{x-r}{u-r} & \text{if } r \leq x \leq u \\ 1 & \text{if } u \leq x \leq U \\ \frac{R-x}{R-U} & \text{if } U \leq x \leq R \\ 0 & \text{otherwise} \end{cases}$$

Proposition 1 (Linear combination of TrFNs [31]) *Let* $\tilde{B} = \sum_{j=1}^n \tilde{u}_j \cdot a_j$ *where* $a_j \in \mathbb{R}$, $a_j \geq 0, j = 1, \ldots, n$, *and let* $\tilde{u}_j, j = 1, \ldots, n$ *be TrFNs defined by* $\tilde{u}_j = (r_j, u_j, U_j, R_j)$. *Then the membership function of* $\tilde{B}$ *is*

$$\mu_{\tilde{B}}(x) = \begin{cases} \frac{x-\mathbf{ra}}{\mathbf{ua}-\mathbf{ra}} & \text{if } \mathbf{ra} \leq x \leq \mathbf{ua} \\ 1 & \text{if } \mathbf{ua} \leq x \leq \mathbf{Ua} \\ \frac{\mathbf{Ra}-x}{\mathbf{Ra}-\mathbf{Ua}} & \text{if } \mathbf{Ua} \leq x \leq \mathbf{Ra} \\ 0 & \text{otherwise} \end{cases}$$

where $\mathbf{r} = (r_1, \ldots, r_n)$, $\mathbf{u} = (u_1, \ldots, u_n)$, $\mathbf{U} = (U_1, \ldots, U_n)$ *and* $\mathbf{R} = (R_1, \ldots, R_n)$. *In other words,* $\tilde{B} = \sum_{j=1}^n \tilde{u}_j \cdot a_j = (\mathbf{ra}, \mathbf{ua}, \mathbf{Ua}, \mathbf{Ra})$.

2.1 Operations with Fuzzy Numbers

In the remainder of this work, only TrFNs will be used exclusively. For this reason and due to space constraints, we only review operations with TrFNs. It is well known that the product and quotient operations with TrFNs do not yield another TrFN. However, only addition, subtraction and product by a scalar are needed for the FLP algorithms implemented here. Hence we focus on those operations.

Proposition 2 *Let* $\tilde{x}_i = (r_i, u_i, U_i, R_i)$, $i = 1, 2$, *two TrFNs. Then*

(i) $\tilde{x}_1 + \tilde{x}_2 = (r_1 + r_2, u_1 + u_2, U_1 + U_2, R_1 + R_2)$
(ii) $\tilde{x}_1 - \tilde{x}_2 = (r_1 - R_2, u_1 - U_2, U_1 - u_2, R_1 - r_2)$
(iii) *If* $a \in \mathbb{R}^+$, $a \cdot \tilde{x}_1 = a \cdot (r_1, u_1, U_1, R_1) = (a \cdot r_1, a \cdot u_1, a \cdot U_1, a \cdot R_1)$
(iv) *If* $a \in \mathbb{R}^-$, $a \cdot \tilde{x}_1 = a \cdot (r_1, u_1, U_1, R_1) = (a \cdot R_1, a \cdot U_1, a \cdot u_1, a \cdot r_1)$.

2.2 Comparison of Fuzzy Numbers

As will be explained later, FLP ultimately requires comparing fuzzy numbers. This is a broad topic by itself, and a lot of proposals have been published; see for instance [16, 24]. Here the method based on ordering functions has been applied, although the code is prepared for other customized comparison functions implemented by the user that can be passed as an argument.

Ordering Functions

Let $\mathcal{F}(\mathbb{R})$ denote the set of fuzzy numbers on $\mathbb{R}$. An ordering (or defuzzification) function is an application $g : \mathcal{F}(\mathbb{R}) \to \mathbb{R}$ so that fuzzy numbers are sorted according to their corresponding defuzzified real numbers. Therefore, given $\tilde{A}, \tilde{B} \in \mathcal{F}(\mathbb{R})$, we consider $\tilde{A} <^g \tilde{B} \Leftrightarrow g(\tilde{A}) < g(\tilde{B})$; $\tilde{A} >^g \tilde{B} \Leftrightarrow g(\tilde{A}) > g(\tilde{B})$; and $\tilde{A} =^g \tilde{B} \Leftrightarrow g(\tilde{A}) = g(\tilde{B})$. Function g is called a linear ordering function if (a) $\forall \tilde{A}, \tilde{B} \in \mathcal{F}(\mathbb{R})$, $g(\tilde{A} + \tilde{B}) = g(\tilde{A}) + g(\tilde{B})$, and (b) $\forall r \in \mathbb{R}, r > 0$ and $\forall \tilde{A} \in \mathcal{F}(\mathbb{R})$, $g(r \cdot \tilde{A}) = r \cdot g(\tilde{A})$.

The following linear ordering functions have been implemented:

1. First Yager's index [39]:

$$g(\tilde{u}) = \frac{\int_S h(z)\mu_{\tilde{u}}(z)dz}{\int_S \mu_{\tilde{u}}(z)dz} \tag{1}$$

   where $S = \text{supp}(\tilde{u})$ and $h(z)$ is a measure of importance of each value z. Taking $h(z) = z$ and assuming TrFNs, this simplifies to

$$g(\tilde{u}) = \frac{1}{3} \frac{(U^2 - u^2) + (R^2 - r^2) + (RU - ru)}{(U - u) + (R - r)} \tag{2}$$

2. Third Yager's index [39]:

$$g(\tilde{u}) = \int_0^1 M(\tilde{u}_\alpha)d\alpha \tag{3}$$

   where $\tilde{u}_\alpha$ is an α-cut of $\tilde{u}$ and $M(\tilde{u}_\alpha)$ the mean value of the elements in $\tilde{u}_\alpha$. When using TrFNs the above expression simplifies to

$$g(\tilde{u}) = \frac{U + u + R + r}{4} \tag{4}$$

3. Adamo relation [1]. For a fixed $\alpha \in [0, 1]$:

$$g_\alpha(\tilde{u}) = \max\{x \, / \, \mu_{\tilde{u}}(x) \geq \alpha\} \tag{5}$$

which for TrFNs simplifies to

$$g_\alpha(\tilde{u}) = R - \alpha(R - U) \qquad (6)$$

4. Average relation [19]. For TrFNs:

$$g_t^\lambda(\tilde{u}) = u - \frac{u - r}{t + 1} + \lambda \left(U - u + \frac{(R - r) - (U - u)}{t + 1} \right), \ \lambda \in [0, 1], \ t \geq 0 \quad (7)$$

where λ represents the degree of optimism to be selected by the decision-maker (larger λ corresponds to a more optimist decision-maker).

3 Fuzzy Linear Programming

A crisp LP problem consists in maximizing/minimizing a function subject to constraints over the variables:

$$\max z = \mathbf{cx}$$
$$s.t. : \mathbf{Ax} \leq \mathbf{b}$$
$$\mathbf{x} \geq \mathbf{0}$$

where $\mathbf{A} \in \mathcal{M}_{mxn}(\mathbb{R})$ is a matrix of real numbers, $\mathbf{c} \in \mathbb{R}^n$ is a cost vector and $\mathbf{b} \in \mathbb{R}^m$ a vector.

In the above formulation, all coefficients are assumed to be perfectly known. However, this is not the case in many real cases of application. There may be uncertainty concerning some coefficients, or they may come from an (approximate) estimation by a human expert/decision maker who will possibly be more confident when expressing her knowledge in linguistic terms of the natural language [40]. Very often, when a person is asked to express her expertise in strictly numerical values, he/she feels like being forced to commit an error, hence the use of natural language, supported by fuzzy numbers to do computations, might be more suitable. Optimization problems with fuzzy quantities were first presented in [3]. Key concepts such as fuzzy constraint and fuzzy goal, which we will explain in detail later in this section, were conceived there.

In the next sections we explain in detail the three FLP models and several solution methods implemented in our package.

3.1 Fuzzy Constraints

We consider the case where the decision maker can accept a violation of the constraints up to a certain degree he/she establishes. This can be formalized for each
constraint as

$$a_i x \leq_f b_i, \quad i = 1, \dots, m$$

and can be modeled using a membership function

$$\mu_i : \mathbb{R} \to [0, 1], \quad \mu_i(x) = \begin{cases} 1 & \text{if } x \leq b_i \\ f_i(x) & \text{if } b_i \leq x \leq b_i + t_i \\ 0 & \text{if } x \geq b_i + t_i \end{cases} \tag{8}$$

where the f_i are continuous, non-increasing functions. Membership functions μ_i capture the fact that the decision maker tolerates a certain degree of violation of each
constraint, up to a value of $b_i + t_i$. For each $x \in \mathbb{R}$, $\mu_i(x)$ stands for the degree of
fulfillment of the i-th constraint for that x. The problem to be solved is

$$\max z = \mathbf{cx}$$
$$s.t. : \mathbf{Ax} \leq_f \mathbf{b}$$
$$\mathbf{x} \geq \mathbf{0}$$

To illustrate the use of the functions implemented to solve this type of problem, we
will use the fuzzy constraints problem shown on the left, which can be transformed
into the problem on the right if the membership functions are assumed linear with
maximum tolerances of 5 and 6.

$$\max z = 3x_1 + x_2 \qquad\qquad\qquad \max z = 3x_1 + x_2$$
$$s.t. : 1.875x_1 - 1.5x_2 \leq_f 4 \quad\Rightarrow\quad s.t. : 1.875x_1 - 1.5x_2 \leq 4 + 5(1 - \alpha)$$
$$4.75x_1 + 2.125x_2 \leq_f 14.5 \quad\Rightarrow\quad 4.75x_1 + 2.125x_2 \leq 14.5 + 6(1 - \alpha)$$
$$x_i \geq 0, i = 1, 2, 3 \qquad\qquad\qquad \alpha_i \geq 0, i = 1, 2, 3$$

The following R commands create the necessary objects:

```
> objective<-c(3, 1)
> A<-matrix(c(1.875, -1.5, 4.75, 2.125), nrow = 2, byrow = T)
> dir = c("<=", "<=") # direction of the inequalities
> b<-c(4, 14.5)
> t<-c(5, 6) # tolerances
```

Different solutions have been proposed. The solution given in [36] generalizes
those given before by [32, 41], which are obtained as particular cases depending on

the value of a parameter. We implement four solution methods, whose names begin with FCLP for Fuzzy Constraints Linear Program

Method 1: Verdegay's Approach

In [36] it was proved, via the representation theorem, that the problem can be solved by solving the following Parametric Linear Programming problem:

$$\max z = \mathbf{cx}$$
$$s.t. : \mathbf{Ax} \leq \mathbf{g}(\alpha)$$
$$\mathbf{x} \geq \mathbf{0}, \alpha \in [0,1]$$

where $\mathbf{g}(\alpha) = (g_1(\alpha), \dots, g_m(\alpha)) \in \mathbb{R}^m$, with $g_i = f_i^{-1}$.
If the f_i are linear, the problem simplifies to

$$\max z = \mathbf{cx}$$
$$s.t. : \mathbf{Ax} \leq \mathbf{b} + \mathbf{t}(1-\alpha)$$
$$\mathbf{x} \geq \mathbf{0}, \alpha \in [0,1]$$

with $\mathbf{t} = (t_1, \dots, t_m) \in \mathbb{R}^m$.

It has been proved [12] that by solving a model with linear f_i, it is possible to obtain the solution to the same fuzzy constraints problem as if it had been modeled with non-linear functions, hence no generality is lost when assuming linear functions for the fuzzy constraints.

Two R functions implement this approach in our package:

- FCLP.fixedBeta which, for a fixed value of α (called β in our function), solves the model. In the example below, $\beta = 0.5$.

```
> FCLP.fixedBeta(objective, A, dir, b, t, beta=0.5, T, T)


[1] "Solution is optimal."
      beta       x1         x2 objective
[1,]   0.5 3.606188 0.1744023  10.99297
```

- FCLP.sampledBeta samples α in the interval $[0,1]$ and solves the model for every sampled value. In the example below we set a step size of 0.25 for sampling α, which yields five α-cuts, for $\alpha = \{0, 0.25, 0.5, 0.75, 1\}$:

```
> FCLP.sampledBeta(objective, A, dir, b, t, T, min=0,
+                    max=1, step=0.25)
       beta        x1         x2 objective
[1,]  0.00 4.315789 0.0000000 12.947368
[2,]  0.25 4.000000 0.0000000 12.000000
[3,]  0.50 3.606188 0.1744023 10.992968
[4,]  0.75 3.164557 0.4556962  9.949367
[5,]  1.00 2.722925 0.7369902  8.905767
```

Method 2: Zimmermann's Approach

In [41, 42] the author discusses the case in which the decision maker is satisfied with a solution that achieves a goal $z_0 \in \mathbb{R}$ for the objective function that, despite not being optimal, minimizes the degree of violation of the constraints. The original formulation is shown on the left. This problem is equivalent to the one on the right, where an additional fuzzy linear constraint has been added. Please notice the new constraint can be thought as embedded in the general expression of the linear constraints, $\mathbf{Ax} \leq_f \mathbf{b}$ if we assume that $\mathbf{A}$ and $\mathbf{b}$ include an extra row for $-\mathbf{c}$ and an element $-z_0$ respectively.

$$
\begin{array}{ccc}
\begin{aligned}
\max \; & \mathbf{cx} \geq_f z_0 \\
s.t. \; : \; & \mathbf{Ax} \leq_f \mathbf{b} \\
& \mathbf{x} \geq 0
\end{aligned}
&\Rightarrow&
\begin{aligned}
\max \; & z = \mathbf{cx} \\
s.t. \; : \; & \mathbf{cx} \geq_f z_0 \\
& \mathbf{Ax} \; \leq_f \mathbf{b}, \mathbf{x} \geq 0
\end{aligned}
\end{array}
$$

When the membership functions of the constraints and the objective are linear the above problem simplifies to

$$
\begin{aligned}
\max \; & \alpha \\
s.t. \; : \; & \mathbf{cx} \geq z_0 - t_0(1 - \alpha) \\
& \mathbf{Ax} \leq \mathbf{b} + \mathbf{t}(1 - \alpha) \\
& \mathbf{x} \geq 0, \alpha \in [0, 1]
\end{aligned}
$$

Two R functions implement this approach:

- `FCLP.classicObjective` for the case that the goal z_0 is crisp.

Goal attainable ($z_0 = 11$):

```
> FCLP.classicObjective(objective, A, dir, b, t, z0=11, TRUE)
```

```
Bound reached, FCLP.fixedBeta with beta = 0.4983154 may
obtain better results.
          beta       x1          x2 objective
[1,] 0.4983154 3.609164 0.1725067        11
```

Goal not attainable ($z_0 = 14$):

```
> FCLP.classicObjective(objective, A, dir, b, t, z0=14, TRUE)
```

```
[1] "Minimal bound not reached."
NULL
```

- `FCLP.fuzzyObjective` for a goal ($z_0 = 14$) on which we admit certain tolerance ($t_0 = 2$).

```
> FCLP.fuzzyObjective(objective, A, dir, b, t, z0=14, t0=2, T)
```

```
Bound reached, FCLP.fixedBeta with beta = 0.1636364 may obtain
better results.
           beta        x1 x2 objective
[1,] 0.1636364 4.109091  0  12.32727
```

Actually `FCLP.fuzzyObjective` generalizes `FCLP.classic Objective`. The latter simply calls the former setting the tolerance $t_0 = 0$.

Method 3: Werners's Approach

Following Zimmermann's proposal, it may occur that the decision maker does not want to provide the goal or the tolerance, or does not have an estimate. Werners [38] proposes two extreme points:

$$Z^0 = \inf\{\max_{x \in X} \mathbf{cx}\}, \qquad Z^1 = \sup\{\max_{x \in X} \mathbf{cx}\} \qquad (9)$$

with $X = \{\mathbf{x} \in \mathbb{R}^n \ / \ \mathbf{Ax} \leq_f \mathbf{b}, \ \mathbf{x} \geq \mathbf{0}\}$. Taking $z_0 = Z^0$ and $t_0 = Z^1 - Z^0$ and assuming linear functions for the constraints and the objective, the new problem to be solved can be formulated as

$$\max \alpha$$
$$s.t. : \mathbf{cx} \geq Z^0 - (Z^1 - Z^0)(1 - \alpha)$$
$$\mathbf{Ax} \leq \mathbf{b} + \mathbf{t}(1 - \alpha)$$
$$\mathbf{x} \geq \mathbf{0}, \quad \alpha \in [0, 1]$$

which is a particularization of Zimmermann's formulation with the aforementioned z_0 and t_0.

This method is implemented by function `FCLP.fuzzyUndefined Objective`, in which the goal and tolerance are estimated and then, `FCLP. fuzzyObjective` is called:

```
> FCLP.fuzzyUndefinedObjective(objective, A, dir, b, t, TRUE)
```

```
[1] "Using bound ="    "12.9473684210526"
[1] "Using tolerance =" "4.04160189503294"
          beta       x1         x2 objective
[1,] 0.5080818 3.591912 0.1834957  10.95923
```

Method 4: Tanaka's Approach

Tanaka [32] proposed normalizing the objective function $z = \mathbf{cx}$. Let M be the optimum when the problem is solved considering crisp constraints. Then

$$f : \mathbb{R}^n \to [0, 1], \quad f(x) = \begin{cases} 1 & \text{if } \mathbf{cx} > M \\ \frac{\mathbf{cx}}{M} & \text{if } \mathbf{cx} \leq M \end{cases}$$

The new problem to be solved is

$$\begin{aligned} \max \; & \alpha \\ s.t. \; : \; & \frac{\mathbf{cx}}{M} \geq \alpha \\ & \mathbf{Ax} \leq \mathbf{b} + \mathbf{t}(1 - \alpha) \\ & \mathbf{x} \geq \mathbf{0}, \; \alpha \in [0, 1] \end{aligned}$$

For implementation purposes, we will do the following modification:

$$\frac{\mathbf{cx}}{M} \geq \alpha \Leftrightarrow \mathbf{cx} \geq M\alpha = M - M(1 - \alpha)$$

Therefore we replace the first constraint above by $\mathbf{cx} \geq M - M(1 - \alpha)$.

This approach is implemented in function FCLP.fuzzyUndefinedNorm Objective. It first computes M and then calls FCLP.fuzzyObjective with $z_0 = t_0 = M$.

```
> FCLP.fuzzyUndefinedNormObjective(objective, A, dir, b, t, TRUE)


[1] "Using bound ="    "12.9473684210526"
[1] "Using tolerance =" "12.9473684210526"
            beta       x1          x2 objective
[1,] 0.7639495 3.139915 0.4713919  9.891136
```

3.2 Fuzzy Costs

FLP with fuzzy costs pose uncertainty in the coefficients of the objective function, modeled as fuzzy numbers. Such problems can be stated as:

$$\begin{aligned} \max \; & z = \tilde{\mathbf{c}}x \\ s.t. \; : \; & \mathbf{Ax} \leq \mathbf{b} \\ & \mathbf{x} \geq \mathbf{0} \end{aligned} \tag{10}$$

where $\mathbf{A} \in \mathcal{M}_{mxn}(\mathbb{R})$ is a real matrix, $\tilde{\mathbf{c}}$ is an n-dimensional vector of fuzzy numbers, and $\mathbf{b} \in \mathbb{R}^m$ is a real vector.

In [10] the costs membership functions are assumed to have the form:

$$\mu_j : \mathbb{R} \rightarrow [0,1], \quad \mu_j(x) = \begin{cases} 0 & \text{if } x \le r_j \text{ or } x \ge R_j \\ h_j(x) & \text{if } r_j \le x \le \underline{c}_j \\ g_j(x) & \text{if } \overline{c}_j \le x \le R_j \\ 1 & \text{if } \underline{c}_j \le x \le \overline{c}_j \end{cases}$$

with h_j and g_j continuous, h_j strictly increasing, g_j strictly decreasing, and such that functions μ_j are continuous.

In order to demonstrate the functions of our package dealing with fuzzy costs, we will use the following example with TrFNs [10, pp. 94, 125]:

$$\max z = (0,2,2,3)x_1 + (1,3,4,5)x_2$$
$$s.t. : x_1 + 3x_2 \le 6$$
$$x_1 + x_2 \le 4$$
$$x_i \ge 0, i = 1,2$$

The following R commands create the necessary objects:

```
> objective<-c(TrapezoidalFuzzyNumber(0,2,2,3), # fuzzy costs
+               TrapezoidalFuzzyNumber(1,3,4,5)) # vector
> A<-matrix(c(1, 3, 1, 1), nrow = 2, byrow=T)
> dir = c("<=", "<=") # direction of the inequalities
> b<-c(6, 4)
```

Four approaches will be presented: three of them are based on the Representation Theorem and the last one, on the comparison of fuzzy numbers.

Method 1: Multi-objective Approach

The results of [13, 36] together with the definition of TrFNs lead us to transform the above problem in the following parametric linear programming problem. Let $X = \{\mathbf{x} \in \mathbb{R}^n \,/\, \mathbf{Ax} \le \mathbf{b}, \, \mathbf{x} \ge \mathbf{0}\}$. Then

$$\max z = \mathbf{cx}$$
$$s.t. : \mathbf{x} \in X$$
$$\Phi(1 - \alpha) \le \mathbf{c} \le \Psi(1 - \alpha) \tag{11}$$
$$\alpha \in [0,1]$$

where $\Phi = (h_1^{-1}, \ldots, h_n^{-1})$ and $\Psi = (g_1^{-1}, \ldots, g_n^{-1})$. Symbols h_j and g_j take part in the definition of the membership functions μ_j of every fuzzy cost $\tilde{c}_j$.

If we fix α and solve the above problem with it, the solution set is the α-cut of the solution fuzzy number, which would be completely defined by all its α-cuts as stated by the Representation Theorem. If we define $\Gamma(1 - \alpha) = \{\mathbf{c} \in \mathbb{R}^n \,/\, c_i \in [\Phi_i(1 - \alpha), \Psi_i(1 - \alpha)]\}$, the new problem constitutes a multi-objective linear programming problem with one objective for each $\mathbf{c} \in \Gamma(1 - \alpha)$. According to [4] the problem is equivalent to the following:

$$\max \{\mathbf{c}^1 \mathbf{x}, \ldots, \mathbf{c}^{2^n} \mathbf{x}\}$$

$$s.t. : \mathbf{Ax} \leq \mathbf{b}$$
$$\mathbf{x} \geq \mathbf{0}$$
$$\mathbf{c}^k \in E(1 - \alpha), \ k = 1, 2, \ldots, 2^n$$
$$\alpha \in [0, 1]$$

where $E(1 - \alpha) \subseteq \Gamma(1 - \alpha)$ is the subset of those vectors whose components are the upper bounds of c_j, i.e., the Cartesian product:

$$E(1 - \alpha) = \prod_{i=1}^{n} \{\Phi_i(1 - \alpha), \Psi_i(1 - \alpha)\}$$

This problem can be solved by any multi-objective linear programming technique. In our code, the objectives have been aggregated using a weighting vector with the same weight for every objective. The objective function thus simplifies to max $\mathbf{c}^1 \mathbf{x} + \cdots + \mathbf{c}^{2^n} \mathbf{x}$ subject to the constraints stated above.

This approach is implemented by function FOLP.multiObj. Since the problem has to be solved for every $\alpha \in [0, 1]$, the function samples α in $[0, 1]$ according to a user-specified step, and solves for each α.

```
> sal<-FOLP.multiObj(objective, A, dir, b, maximum=TRUE, min=0,
+                          max=1, step=0.25)
> sal


        alpha x1 x2 objective
[1,] 0        3  1  ?
[2,] 0.25     3  1  ?
[3,] 0.5      3  1  ?
[4,] 0.75     3  1  ?
[5,] 1        3  1  ?


> sal[,"objective"] # Display the objective column properly


[[1]]
Trapezoidal fuzzy number with:
    support=[1,14],
        core=[9,10].
... # output omitted for elements 2, 3, 4 and 5
```

Method 2: Interval Arithmetic Approach

Expression (11) can be viewed as a linear programming problem in which every coefficient of the objective function takes values in an interval. Therefore, the problem can be solved resorting to interval arithmetic and relations $\leq_l, \leq_c, \leq_{lc}$ introduced in [2, 21].

Definition 3 Let $A = [a^l, a^u] =< a^c, a^w >$ and $B = [b^l, b^u] =< b^c, b^w >$ be two intervals, where the $< \cdot, \cdot >$ are based on the center c and width w.

- $A \leq_l B$ if $a^l \leq b^l$ and $a^u \leq b^u$
- $A \leq_c B$ if $a^c \leq b^c$ and $a^w \geq b^w$.
- $A \leq_{lc} B$ if $a^l \leq b^l$ and $a^c \leq b^c$

- $A <_l B$ if $A \leq_l B$ and $A \neq B$
- $A <_c B$ if $A \leq_c B$ and $A \neq B$
- $A <_{lc} B$ if $A \leq_{lc} B$ and $A \neq B$

For every $\mathbf{x} \in X, \alpha \in [0, 1]$, define the intervals

$$I_j(\alpha) = [\Phi_j(1 - \alpha), \Psi_j(1 - \alpha)] \qquad \text{and} \qquad z(\mathbf{x}, \alpha) = \sum_{j=1}^{n} x_j I_j(\alpha)$$

For each α, a solution $\mathbf{x}^*$ to the problem is one whose associated interval $z(\mathbf{x}^*, \alpha)$ is non-dominated, i.e. $\mathbf{x}^* \in X$ such that $\nexists \mathbf{x}' \in X : z(\mathbf{x}^*, \alpha) \leq_{lc} z(\mathbf{x}', \alpha)$. Since $\mathbf{x}^*$ does not have to be unique, we can define the set

$$S(1 - \alpha) = \{\mathbf{x} \in X \, / \, \nexists \mathbf{x}' \in X : z(\mathbf{x}, \alpha) \leq_{ic} z(\mathbf{x}', \alpha)\}$$

These sets are the α-cuts of the fuzzy solution which, according to the Representation Theorem, would yield the solution fuzzy number $\tilde{S} = \bigcup_\alpha \alpha S(1 - \alpha)$.

For a fixed α, the problem of finding solutions whose associated intervals are non-dominated can be formulated as the following bi-objective problem.

$$\max\{z(\alpha) = (z^i(\mathbf{x}, \alpha), z^c(\mathbf{x}, \alpha)) : \mathbf{x} \in X\}$$

According to [20] this problem can be solved using weights. Let $\beta_1, \beta_2 \in [0, 1] :$ $\beta_1 + \beta_2 = 1$. The problem can be reformulated as:

$$\max\{z(\alpha) = \beta_1 z^i(\mathbf{x}, \alpha) + \beta_2 z^c(\mathbf{x}, \alpha) : \mathbf{x} \in X\}$$

This approach is implemented by function FOLP.interv which receives the problem data and weight β_1 (note β_2 can be automatically computed as 1 - β_1). The function performs samples α in [0, 1] with the user-specified step size. A private function computes $z(\alpha)$ from the fuzzy coefficients.

```
> sal<-FOLP.interv(objective, A, dir, b, maximum=TRUE, w1=0.7,
+                  min=0, max=1, step=0.25)
> sal
```

The structure of this variable is the same as in the previous section.

Method 3: Stratified Piecewise Reduction

In [27] the fuzzy cost problem is approached by modeling the uncertainty of the coefficients using embedded intervals, each with an associated possibility degree:

$$\tilde{c}_j = \{[r_j^{(k)}, R_j^{(k)}] \mathbin/ \alpha^{(k)}; k = 1, \dots, p\}$$

For a given $\alpha \in [0, 1]$, consider the intervals obtained from the α-cuts of the fuzzy coefficients. With a slight abuse of notation (and omitting the α that has been fixed), we will write $\tilde{c}_j = [r_j, R_j]$.

Let $\mathbf{r} = (r_1, \dots, r_n)$, $\mathbf{R} = (R_1, \dots, R_n)$, and consider the LP problems

$$\max z = \mathbf{rx} \qquad\qquad \max z = \mathbf{Rx}$$
$$s.t. : \mathbf{Ax} \le \mathbf{b} \qquad\qquad s.t. : \mathbf{Ax} \le \mathbf{b}$$
$$\mathbf{x} \ge \mathbf{0} \qquad\qquad\qquad \mathbf{x} \ge \mathbf{0}$$

Let $\mathbf{x}_\mathbf{r}^*$ and $\mathbf{x}_\mathbf{R}^*$ be their respective solutions.

Let $z_\mathbf{r}^* = \mathbf{rx}_\mathbf{r}^*$ and $z_\mathbf{R}^* = \mathbf{Rx}_\mathbf{R}^*$ be the optimal solutions of the objective functions, and let $z_\mathbf{r}' = \mathbf{rx}_\mathbf{R}^*$ and $z_\mathbf{R}' = \mathbf{Rx}_\mathbf{r}^*$. Clearly $z_\mathbf{r}^* \ge z_\mathbf{r}'$ and $z_\mathbf{R}^* \ge z_\mathbf{R}'$.

The problem can be solved using the auxiliary problem

$$\max \lambda$$
$$s.t. : \quad \frac{\mathbf{rx} - z_\mathbf{r}'}{z_\mathbf{r}^* - z_\mathbf{r}'} \ge \lambda$$
$$\frac{\mathbf{Rx} - z_\mathbf{R}'}{z_\mathbf{R}^* - z_\mathbf{R}'} \ge \lambda$$
$$\mathbf{Ax} \le \mathbf{b}, \mathbf{x} \ge \mathbf{0}, \lambda \ge 0$$

After solving the above problems for different values of α, the solution to the original fuzzy costs problem can be found as the intersection of the solutions of the auxiliary problems.

Function FOLP.strat computes the values $z_\mathbf{r}^*, z_\mathbf{r}', z_\mathbf{R}^*$ and $z_\mathbf{R}'$ by solving the corresponding LP problems, and then solves the original problem. This has to be done separately for each value of α, therefore FOLP.strat samples $\alpha \in [0, 1]$ with a user-specified step size.

```
> sal <- FOLP.strat(objective, A, dir, b, maximum=TRUE, min=0,
+                   max=0.4, step=0.05)
> sal
```

```
      alpha x1   x2    lambda objective
```

```
[1,] 0       1.5 1.5 0.5      ?
[2,] 0.05    1.5 1.5 0.5      ?
[3,] 0.1     1.5 1.5 0.5      ?
[4,] 0.15    1.5 1.5 0.5      ?
[5,] 0.2     1.5 1.5 0.5      ?
[6,] 0.25    3   1   1        ?
[7,] 0.3     NA  NA  NA       NA
[8,] 0.35    NA  NA  NA       NA
[9,] 0.4     NA  NA  NA       NA
```

```
> sal[,"objective"] # display the objective column properly
```

```
[[1]]
Trapezoidal fuzzy number with:
   support=[1.5,12],
      core=[7.5,9].
... # output omitted for list elements 2 to 9
```

Method 4: Ordering Functions

The problem (10) can be transformed into a crisp one by using a linear ordering function $g : \mathcal{F}(\mathbb{R}) \to \mathbb{R}$, so that the fuzzy objective function is replaced by max $z = g(\tilde{c}_1)x_1 + \cdots + g(\tilde{c}_n)x_n$, subject to the same crisp constraints.

Function FOLP.ordFun implements this approach. It receives the problem data and a string indicating the ordering function to be used (argument ordf). This can be one of the four built-in functions, namely "Yager1" for the first Yager's index (Eq. 2), "Yager3" for the third (Eq. 4), "Adamo" for the Adamo relation (Eq. 6), and "Average" for the average index (Eq. 7). Some of them require additional arguments, as described in detail in the package documentation. If a user-defined custom linear function is to be used, the string should be "Custom". The custom function is passed to FOLP.ordFun in the FUN argument. The custom function must accept at least one argument of class FuzzyNumber, and may also accept additional arguments that must be named when passing them to FOLP.ordFun. It is the user's responsibility to give them names that do not interfere with existing ones, and to care that the function is linear.

Example call using first Yager's index:

```
> sal<-FOLP.ordFun(objective, A, dir, b, maximum=TRUE,
+                  ordf="Yager1")
```

For Adamo's index, which requires an additional parameter α:

```
> sal<-FOLP.ordFun(objective, A, dir, b, maximum=TRUE,
+                  ordf="Adamo", alpha=0.5)
```

For a custom function that computes the mean of the core multiplied by another real number:

```
> custom.f <- function(tfn,a){ a * mean(core(tfn)) }
> sal<-FOLP.ordFun(objective, A, dir, b, TRUE, "Custom",
+                  FUN=custom.f, a=2)
```

3.3 General Model

The most general setting is that with fuzzy costs, fuzzy coefficients in the technology matrix, and fuzzy constraints that can be violated up to a certain degree. Calling m to the number of constraints, it can be formalized as the problem on the left. This formulation can be transformed into the problem on the right, according to the Representation Theorem and assuming that the decision maker agrees with considering the same degree of satisfaction both in the fuzzy costs and in the fuzzy technological matrix.[5]

$$\begin{array}{ll} \max z = \tilde{c}x \\ s.t. : \tilde{a}_i x \leq_f \tilde{b}_i, \\ \quad\quad x \geq 0 \end{array} \quad\Rightarrow\quad \begin{array}{ll} \max z = \tilde{c}x \\ s.t. : \tilde{a}_i x \leq \tilde{b}_i + \tilde{t}_i(1-\alpha), \quad i = 1, \ldots, m \\ \quad\quad x \geq 0 \end{array}$$

where $\tilde{a}_i$ and $\mathbf{b}_i (i = 1, \ldots, m)$ are n-dimensional vectors of fuzzy numbers, $\tilde{c}$ is another n-dimensional vector of fuzzy numbers, and $\tilde{t}_i$ is the fuzzy tolerance admitted for violating the i-th constraint.

Let g_1 and g_2 be two linear ordering functions for the objective and for the constraints, respectively. With them, and because g_1 and g_2 are linear, the problem can be defuzzified to obtain the following crisp LP problem:

$$\max z = g_1(\tilde{c})x$$
$$s.t. : g_2(\tilde{a}_i)x \leq g_2(\tilde{b}_i) + g_2(\tilde{t}_i)(1-\alpha), \quad i = 1, \ldots, m$$
$$x \geq 0$$

The function implementing this approach is called GFLP. It receives the two linear ordering functions to be used (one for the objective function and the other for the constraints). They must be one of the functions described in Sect. 3.2. Since the problem has to be solved for a fixed α and then the Representation Theorem is used, the function samples $\alpha \in [0, 1]$ with a user-specified step size.

---

[5]Otherwise, different α- and β-cuts should be needed and the problem would become more difficult.

The function will be demonstrated with the following example:

$$\max z = (1, 3, 4, 5)x_1 + (0, 1, 1, 2)x_2$$
$$s.t. : (0, 2, 2, 3.5)x_1 + (0, 1, 1, 4)x_2 \leq (2, 2, 2, 3) + (1, 2, 2, 3)(1 - \alpha)$$
$$(3, 5, 5, 6)x_1 + (1.5, 2, 2, 3)x_2 \leq 12$$
$$x_1 \geq 0, x_2 \geq 0$$

The R commands below create the necessary objects:

```
> objective<-c(TrapezoidalFuzzyNumber(1,3,4,5),
+                TrapezoidalFuzzyNumber(0,1,1,2))
> A<-matrix(c(TrapezoidalFuzzyNumber(0,2,2,3.5),
+              TrapezoidalFuzzyNumber(3,5,5,6),
+              TrapezoidalFuzzyNumber(0,1,1,4),
+              TrapezoidalFuzzyNumber(1.5,2,2,3)), nrow= 2)
> dir = c("<=", "<=")
> b<-c(TrapezoidalFuzzyNumber(2,2,2,3), 12)
> t<-c(TrapezoidalFuzzyNumber(1,2,2,3),0)
```

The example employs the *average index* (which receives two additional parameters λ and t) for the objective function, and Adamo for the constraints. As the latter only requires one parameter, it can be passed directly without using a tagged vector.

```
> sal<-GFLP(objective, A, dir, b, t, TRUE, "Average",
+              ordf_obj_param=c(lambda=0.5, t=3),
+              ordf_res="Adamo", ordf_res_param = 0.5)
> sal


      beta x1          x2 objective
[1,]  0    1.818182    0  ?
[2,]  0.25 1.590909    0  ?
[3,]  0.5  1.363636    0  ?
[4,]  0.75 1.136364    0  ?
[5,]  1    0.9090909   0  ?


> sal[,"objective"]


[[1]]
Trapezoidal fuzzy number with:
   support=[1.81818,9.09091],
      core=[5.45455,7.27273].
# ... output omitted for elements 2 to 5
```

4 Conclusions and Further Work

An R package for solving FLP problems has been presented and demonstrated in simple use cases. It can deal with fuzzy constraints, fuzzy costs and a fuzzy technology matrix, and provides specific functions for solving each type of problem. The computations are done with TrFNs as they ease the application of several theoretical results. Our code relies on packages FuzzyNumbers for creating and working with TrFNs, and ROI for solving the crisp LP problems in which the FLP problems are transformed. The package, called FuzzyLP, can be downloaded from CRAN, the R centralized repository.

To the best of our knowledge, this is the first open-source implementation of FLP solving methods, and possibly the only one available in a modern language, as R is. It has been developed as a library of functions, which broadens its usability. Nevertheless, much more can still be done in this direction, such as incorporating more solving techniques for other types of FLP problems, and spanning the functionality to fuzzy non-linear optimization, such as Fuzzy Quadratic Programming (FQP).

Acknowledgments David A. Pelta and José Luis Verdegay want to acknowledge Ronald Yager for his support, help and sincere friendship. This work is supported by projects TIN2011-27696-C02-01 from the Spanish Ministry of Science and Innovation, P11-TIC-8001 from the Andalusian Government, and FEDER funds. The first author acknowledges an FPU scholarship from the Spanish Ministry of Education.

References

1. Adamo, J.M.: Fuzzy decision trees. Fuzzy Sets Syst. **4**, 207–219 (1980)
2. Alefeld, G., Herzberger, J.: Introduction to Interval Computations. Academic Press, NY (1984)
3. Bellman, R., Zadeh, L.: Decision making in a fuzzy environment. Management Science **17**(B) 4 141–164 (1970)
4. Bitran, G.: Linear multiple objective problems with interval coefficients. Manage. Sci. **26**(7), 694–706 (1985)
5. Brito, J., Martinez, F., Moreno, J., Verdegay, J.: Fuzzy optimization for distribution of frozen food with imprecise times. Fuzzy Optim. Decis. Making **11**(3), 337–349 (2012)
6. Cadenas, J., Liern, V., Sala, R., Verdegay, J.: Fuzzy Optimization, chap. Fuzzy Linear Programming in Practice: An Application to the Spanish Football League, pp. 503–528. Studies in Fuzziness and Soft Computing. Springer (2010)
7. Cadenas, J., Pelta, D., Pelta, H., Verdegay, J.: Application of fuzzy optimization to diet problems in Argentinean farms. Eur. J. Oper. Res. **158**, 218–228 (2004)
8. Cadenas, J., Pelta, D., Verdegay, J.: Introducing SACRA: a decision support system for the construction of cattle diets. In: Applied Decision Support with Soft Computing, pp. 391–401. Springer (2003)
9. Cadenas, J.M., Verdegay, J.L.: PROBO: an interactive system in fuzzy linear programming. Fuzzy Sets Syst. **76**, 319–332 (1995)
10. Cadenas, J.M., Verdegay, J.L.: Optimization Models with Imprecise Data (in Spanish). Servicio de Publicaciones, University of Murcia (1999)
11. Dantzig, G.B.: Origins of the simplex method. Technical Report SOL 87-5, Department of Operations Research, Stanford University, Stanford, CA (1987)

12. Delgado, M., Herrera, F., Verdegay, J.L., Vila, M.A.: Post-optimality analysis on the membership function of a fuzzy linear programming problem. Fuzzy Sets Syst. **53**, 289–297 (1993)
13. Delgado, M., Verdegay, J.L., Vila, M.A.: Imprecise costs in mathematical programming problems. Control Cybern. **16**(2), 113–121 (1987)
14. Dubois, D., Prade, H.: Operations on fuzzy numbers. Int. J. Syst. Sci. **9**, 613–626 (1978)
15. Dubois, D., Prade, H.: Fuzzy Sets and Systems. Theory and Applications. Academic Press (1980)
16. Dubois, D., Prade, H.: Ranking fuzzy numbers in the setting of possibility theory. Inf. Sci. **30**(3), 183–224 (1983)
17. Fedrizzi, M., Kacprzyk, J., Verdegay, J.: A survey of fuzzy optimization and mathematical programming. In: Fedrizzi, M., Kacprzyk, J., Roubens, M. (eds.) Interactive Fuzzy Optimization. Lecture Notes in Economics and Mathematical Systems, vol. 368, pp. 15–28. Springer, Berlin (1991)
18. Gagolewski, M., Caha, J.: FuzzyNumbers Package: Tools to Deal with Fuzzy Numbers in R (2015). http://FuzzyNumbers.rexamine.com/
19. González, A.: A study of the ranking function approach through mean values. Fuzzy Sets Syst. **35**, 29–41 (1990)
20. Kornbluth, J.S.H., Steuer, R.E.: Multiple objective linear fractional programming. Manage. Sci. **27**(9), 1024–1039 (1981)
21. Moore, R.E.: Methods and Applications of Interval Analysis. SIAM Studies in Applied and Numerical Mathematics, book 2. SIAM (1979)
22. Negoita, C.V., Ralescu, D.A.: Applications of Fuzzy Sets to Systems Analysis. Wiley (1975)
23. Peidro, D., Mula, J., Poler, R., Verdegay, J.L.: Fuzzy optimization for supply chain planning under supply, demand and process uncertainties. Fuzzy Sets Syst. **160**(18), 2640–2657 (2009)
24. Prade, H., Yager, R.R., Dubois, D. (eds.): Readings in Fuzzy Sets for Intelligent Systems. Morgan Kaufmann Publishers (1993)
25. R Core Team: R: A Language and Environment for Statistical Computing. R Foundation for Statistical Computing. Vienna (2013). http://www.R-project.org/
26. Rommelfanger, H.: Fuzzy linear programming and applications. Eur. J. Oper. Res. **92**, 512–527 (1996)
27. Rommelfanger, H., Hanuscheck, R., Wolf, J.: Linear programming with fuzzy objectives. Fuzzy Sets Syst. **29**, 31–48 (1989)
28. Rosenthal, R.E.: GAMS: A User's Guide. GAMS Development Corporation, Washington DC (2014)
29. Sadeghi, M., Hosseini, H.M.: Evaluation of fuzzy linear programming application in energy models. Int. J. Energy Optim. Eng. **2**(1), 50–59 (2013)
30. Silva, R.C., Cruz, C., Verdegay, J.L., Yamakami, A.: A survey of fuzzy convex programming models. In: Lodwick, W.A., Kacprzyk, J. (eds.) Fuzzy Optimization, Studies in Fuzziness and Soft Computing, vol. 254, pp. 127–143. Springer, Berlin (2010)
31. Tanaka, H., Ichihashi, H., Asai, F.: A formulation of fuzzy linear programming problems based a comparison of fuzzy numbers. Control Cybern. **13**, 185–194 (1984)
32. Tanaka, H., Okuda, T., Asai, K.: On fuzzy mathematical programming. J. Cybern. **3**(4), 37–46 (1974)
33. Theussl, S., Meyer, D., Hornik, K.: Many Solvers, One Interface: ROI, the R Optimization Infrastructure Package. In: useR! conference, p. 161 (2010). http://www.r-project.org/conferences/useR-2010/abstracts/_Abstracts.pdf
34. Tsakiris, G., Spiliotis, M.: Fuzzy linear programming for problems of water allocation under uncertainty. Eur. Water **7–8**, 25–37 (2004)
35. Vasant, P.M.: Application of fuzzy linear programming in production planning. Fuzzy Optim. Decis. Making **2**(3), 229–241 (2003)
36. Verdegay, J.L.: Fuzzy mathematical programming. In: Gupta, M.M., Sánchez, E. (eds.) Fuzzy Information and Decision Processes, pp. 231–237 (1982)
37. Venables, W.N., Smith, D.M., R Core Team: An Introduction to R, version 3.1.2. R Foundation for Statistical Computing. Vienna (2014). http://cran.r-project.org/doc/manuals/R-intro.pdf

38. Werners, B.: An interactive fuzzy programming system. Fuzzy Sets Syst. **23**, 131–147 (1987)
39. Yager, R.R.: Ranking fuzzy subsets over the unit interval. In: Proceedings of of the IEEE Conference on Decision and Control, pp. 1435–1437 (1978)
40. Zadeh, L.: The concept of a linguistic variable and its applications to approximate reasoning, part I, II and III. Inf. Sci. **8**, 199–249, **8**, 301–357, **9**, 43–80 (1975)
41. Zimmermann, H.J.: Description and optimization of fuzzy systems. Int. J. Gen. Syst. **2**, 209–215 (1976)
42. Zimmermann, H.J.: Fuzzy programming and linear programming with several objective functions. Fuzzy Sets Syst. **1**(1), 45–55 (1978)

Part III
Some Bibliometric Remarks

A Bibliometric Analysis of the Publications of Ronald R. Yager

José M. Merigó, Anna M. Gil-Lafuente and Janusz Kacprzyk

Abstract This study presents a bibliometric analysis of the publications of Ronald R. Yager available in Web of Science. Currently Professor Yager has more than 500 publications in this database. He is recognized as one of the most influential authors in the World in Computer Science. The bibliometric review considers a wide range of issues including a specific analysis of his publications, collaborators and citations. The VOS viewer software is used to visualize his publication and citation network though bibliographic coupling and co-citation analysis. The results clearly show his strong influence in Computer Science although it also shows a strong influence in Engineering and Applied Mathematics.

Keywords Ronald R. Yager · Bibliometric analysis · Web of Science · VOS viewer

1 Introduction

Ronald R. Yager is one of the most influential authors in the world in computer science. He has received many distinctions, exemplified by the recently received Rosenblatt Medal from the IEEE. He has also been included in the

J.M. Merigó (✉)
Department of Management Control and Information Systems,
School of Economics and Business, University of Chile,
Av. Diagonal Paraguay 257, 8330015 Santiago, Chile
e-mail: jmerigo@fen.uchile.cl

A.M. Gil-Lafuente
Department of Business Administration, University of Barcelona, Av. Diagonal 690,
08034 Barcelona, Spain
e-mail: amgil@ub.edu

J. Kacprzyk
Systems Research Institute, Polish Academy of Sciences, Newelska 6 Street,
01-447 Warsaw, Poland
e-mail: Janusz.Kacprzyk@ibspan.waw.pl

© Springer International Publishing Switzerland 2017
J. Kacprzyk et al. (eds.), *Granular, Soft and Fuzzy Approaches*
for Intelligent Systems, Studies in Fuzziness and Soft Computing 344,
DOI 10.1007/978-3-319-40314-4_12

233

Thomson/Reuters list of Most Influential Authors in Computer Science in 2001. He is a Fellow of the New York Academy of Sciences, the International Fuzzy Systems Association (IFSA) and the Institute of Electrical and Electronics Engineers (IEEE), the largest professional organization of this kind in the world. He is on the editorial board of many leading journals including IEEE Transactions on Fuzzy Systems, Fuzzy Sets and Systems, International Journal of Approximate Reasoning, International Journal of General Systems and International Journal of Uncertainty, Fuzziness and Knowledge-Based Systems. Moreover, he is the editor-in-chief of the International Journal of Intelligent Systems. He has published an extremely huge number of publications. Including conference proceedings and related material, he has published more than 1000 publications. Currently, he has almost 50.000 citations according to Google Scholar and the Hirsch [6] h-index above 100. Through this index, he has been included in a very selective list that includes all the authors from all-time and all sciences that currently have an h-index of 100 or more. This list currently includes only 1040 authors (http://www.webometrics.info/en/node/58).

Bibliometrics is a research field that studies the bibliographic material quantitatively [3]. It provides general overviews of a research variable including topics, journals, universities, authors, and countries. In the literature, there are many studies that provide bibliometric overviews in a wide range of issues including topics [8], journals [4, 8, 9] and countries [2].

The aim of this study is to provide a bibliometric overview of the publications of Ronald R. Yager in the Web of Science in order to see his research network. Note that the Web of Science is a database that includes those journals that are usually regarded as the most influential ones. The results clearly show that Yager is one of the World leading authors in Computer Science.

The rest of the paper is organized as follows. Section 2 briefly reviews the methods used in the paper. Section 3 presents the publication and citation structure of Yager and Sect. 4 develops a bibliographic coupling and co-citation analysis of his publications. Section 5 summarizes the main conclusions and findings of the paper.

2   Bibliometric Methods

Bibliometric studies provide a general overview of a research variable such as a country or a journal. In this study, the focus is on an individual author analysis. This approach is useful to analyse deeply leading researchers that have a huge number of publications in order to see their publication and citation structure. This article analyses a wide range of bibliometric indicators [1] including the total number of publications and citations, cites per paper, the h-index [6], and the citing articles. Moreover, we also use the VOS viewer software [11] to visualize the results through bibliographic coupling [7] and co-citation [10]. Note that in this case, bibliographic coupling occurs when two documents of Yager cite the same third document. Co-citation appears when two documents receive a citation by the same third study of Yager.

The search process uses the Web of Science database. Web of Science currently includes more than 15 thousand journals and more than 50 million documents. In order to develop the search process, we searched for "Yager RR" by "Author" in the Web of Science Core Collection, which is usually recognized as the most influential database because it strictly includes the highest-quality material. The search process was carried out the 28th of March, 2016.

Currently, Ronald R. Yager has published 585 publications available in Web of Science Core Collection. Note that by including other materials not available in the WoS database; Yager has more than 1000 publications. Most of them are listed at: http://scholar.google.com/citations?user=uAsllJMAAAAJ&hl=en&oi=ao.

These 585 documents are divided in 418 articles, 141 proceedings articles, 19 letters, 11 editorials, 5 notes and 1 review. When looking at the citation report of the Web of Science, we see that Yager has received 17523 citations with a ratio of 29.95 cites per paper. He has an h-index of 61 and has 9584 citing articles.

3   Publication and Citation Structure of Ronald R. Yager

This section analyses the 585 publications of Ronald R. Yager in Web of Science in order to identify its main research profile and connections. Figure 1 presents the number of documents he has published annually.

During the last twenty years, Yager has published an average of 15–20 articles each year in sources indexed in Web of Science. Next, let us look into the citations he has received annually. Figure 2 presents the number of citations received in each year.

As we can see, the number of citations received has increased significantly throughout time. This is explained because his research has become very popular in the scientific community but also due to some other general factors that has affected science throughout time. During the last years, research has increased a lot due to an

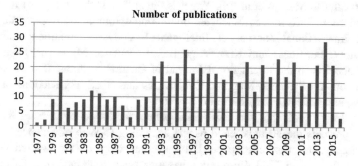

Fig. 1 Number of annual publications

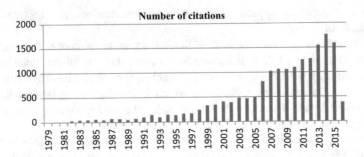

Fig. 2 Number of annual citations

increase in the knowledge economy that invests more and more on people involved in research. During the last decades, this strong increase was seen in Europe but also in many developing countries that currently are growing a lot. Additionally, Web of Science has increased the number of journals covered in the database which implies a higher volume of publications and citations. During the last years, Professor Yager is receiving more than one thousand citations each year and between 2013 and 2015, the number of citations is above 1500 and the trend is that the citations will continue increasing.

Ronald R. Yager has published in many leading journals. Table 1 presents the journals where Yager has published most of his papers. Note that some additional indicators are also shown in the table including the citations, h-index and cites per paper that the publications in this journal has obtained.

Fuzzy Sets and Systems is the journal where Yager has published the highest number of articles. However, the journal with the highest number of citations is the IEEE Transactions on Systems, Man and Cybernetics, mainly because Yager published his seminal paper on OWA operators in this journal. Moreover, note that the journal was divided in three parts in 1996. Thus, if we sum all the publications Yager has published in all the different names of the journal, he would have 51 articles and more than 4500 citations. He also has a significant number of publications in Information Sciences. Currently, he has published more than twenty papers in seven journals and more than ten in thirteen journals.

According to Merigó et al. [8], Yager is the author with the highest number of publications in Information Sciences, International Journal of Approximate Reasoning, International Journal of Intelligent Systems, International Journal of Uncertainty, Fuzziness and Knowledge Based Systems. Moreover, he is in the second position in the International Journal of General Systems, third in all IEEE Transactions on Systems, Man and Cybernetics and IEEE Transactions on Fuzzy Systems, and fifth in Fuzzy Sets and Systems.

Next, let us look into the most cited papers published by Ronald R. Yager. Table 2 presents the 30 most cited papers according to Web of Science.

Obviously, the most cited paper is his seminal work on the ordered weighted average published in 1988. Currently it has more than 2600 citations. This paper is

Table 1 Publications of Ronald R. Yager classified by journals

Journals	Abbrev.	TP	TC	TC/TP	H
Fuzzy Sets and Systems	FSS	59	3355	56.86	24
Information Sciences	IS	49	2140	43.67	21
IEEE Trans. Fuzzy Systems	TFS	40	652	16.30	13
Int. J. Intelligent Systems	IJIS	37	1147	31.00	16
Int. J. General Systems	IJGS	34	1604	47.18	17
IEEE Trans. Systems, Man and Cybernetics[a]	SMC	25	3435	137.40	13
Int. J. Man-Machine Studies	IJMMS	22	718	32.64	11
Int. J. Uncert. Fuzziness Knowledge-Based Syst.	IJUFK	18	446	24.78	7
Int. J. Approximate Reasoning	IJAR	17	778	45.76	11
IEEE Trans. SMC—Part B	SMCB	17	1035	60.88	13
Kybernetes	KYB	14	225	16.07	8
Cybernetics and Systems[b]	CS[b]	13	214	16.46	6
Fuzzy Optimization and Decision Making	FODM	12	292	24.33	7
Soft Computing	SC	7	129	18.43	3
Knowledge-Based Systems	KBS	5	24	4.80	2
J. Intelligent and Fuzzy Systems	JIFS	5	41	8.20	2
IEEE Trans. SMC—Part A	SMCA	5	155	31.00	5
Information Fusion	IF	4	15	3.75	2
4 other		3	–	–	–
7 other		2	–	–	–

[a]Note that in 1996 the IEEE-TSMC was divided in part A, B and C
[b]Between 1971 and 1979, the official name was Journal of Cybernetics

the 96th most cited paper of all-time in Computer Science of more than 2 million publications. It is the second most cited paper in the World in computer science published in 1988 of more than 15000 publications. If focusing on the Web of Science category of Computer Science—Cybernetics, it is the 8th most cited paper of all-time in this category of more than 100000 publications.

His second most cited paper was published in 1993 with 579 citations [13]. Most of his key publications are related to the OWA operators [5, 12, 14]. Currently, he has nine papers with more than 300 citations. Regarding co-authors, it is worth noting his collaboration with Dimitar P. Filev that has provided five of his thirty most cited papers. Additionally, Zeshui Xu has co-authored three papers in the list, Alexander Rybalov has two and Janos C. Fodor, Krassimir Atanassov and Gabriela Pasi, one.

Now, let us analyze the publications according to some different variables. First, Table 3 presents the publications according to the categories and research areas where the journals are usually classified.

As we can see, most of his publications are in the fields of Computer Science. Particularly, Artificial Intelligence is the most influential category with more than 300 publications.

Table 2 30 most cited papers in Web of Science published by Ronald R. Yager

Title	Co-author	J	Y	TC
1. On OWA aggregation operators for multi-criteria decision-making		TSMC	1988	2638
2. Families of OWA operators		FSS	1993	579
3. A procedure for ordering fuzzy subsets of the unit interval		IS	1981	497
4. Quantifier guided aggregation using OWA operators		IJIS	1996	462
5. Some geometric aggregation operators based on intuitionistic fuzzy sets	Xu, ZS	IJGS	2006	421
6. On the Dempster-Shafer framework and new combination rules		IS	1987	399
7. Induced ordered weighted averaging operators	Filev, DP	SMCB	1999	394
8. On a general class of fuzzy connectives		FSS	1980	377
9. Uninorm aggregation operators	Rybalov, A	FSS	1996	323
10. Approximate clustering via the mountain method	Filev, DP	TSMC	1994	273
11. Structure of uninorms	Fodor, JC; Rybalov, A	IJUFK	1997	265
12. On the issue of obtaining OWA operator weights	Filev, DP	FSS	1998	253
13. Dynamic intuitionistic fuzzy multi-attribute decision making	Xu, ZS	IJAR	2008	200
14. Measure of fuzziness and negation. Part 1		IJGS	1979	187
15. On the theory of bags		IJGS	1986	186
16. Entropy and specificity in a mathematical theory of evidence		IJGS	1983	180
17. A new approach to the summarization of data		IS	1982	177
18. Multiple objective decision-making using fuzzy sets		IJMS	1977	172
19. Induced aggregation operators		FSS	2003	164
20. OWA aggregation over a continuous interval argument with applications to decision making		SMCB	2004	163
21. Connectives and quantifiers in fuzzy sets		FSS	1991	159
22. An approach to ordinal decision-making		IJAR	1995	155
23. Intuitionistic fuzzy interpretations of multi-criteria multi-person and multi-measurement tool decision making	Atanassov, K; Pasi, G	IJSS	2005	144
24. A generalized defuzzification method via BAD distributions	Filev, DP	IJIS	1991	116
25. Analytic properties of maximum entropy OWA operators	Filev, DP	IS	1995	114
26. Aggregation operators and fuzzy systems modeling		FSS	1994	111
27. Centered OWA operators		SC	2007	110
28. A characterization of the extension principle		FSS	1986	107

(continued)

Table 2 (continued)

Title	Co-author	J	Y	TC
29. Intuitionistic and interval-valued intutionistic fuzzy preference relations and their measures of similarity for the evaluation of agreement within a group	Xu, ZS	FODM	2009	105
30. Applications and extensions of OWA aggregations		IJMS	1992	105

Abbreviations are available in Table 1 except for: J = Journal; Y = Year

Table 3 Publications distributed by WoS categories and research areas

WoS categories		Research areas	
CS—Artificial Intelligence	302	Computer Science (CS)	547
CS—Theory & Methods	133	Engineering	174
CS—Cybernetics	108	Mathematics	76
Eng. Electrical & Electronics	106	Automation & Control Syst	34
CS—Information Systems	81	Operations Res. & Manag. Sci	30
Mathematics, Applied	73	Psychology	24
Statistics and Probability	61	Business and Economics	11
Ergonomics	58	Imaging Sci Photograph Tech	6
Automation and Control Syst	34	Information Sci & Library Sci.	6
Operations Res. and Manag. Sci	30	Sci Techn Other Topics	4
CS—Interdisciplinary	27	Optics	4
Psychology	23	3 other	3
Management	8	2 other	2
CS—Software Engineering	7	12 other	1
Imaging Sci Photographic Tech	6		
Engineering, Multidisciplinary	6		
Information Sci and Library Sci.	5		
Optics	4		
Logic	4		
5 other	3		
4 other	2		
19 other	1		

Next, let us focus on the main collaborators of Ronald R. Yager. Table 4 presents the collaborators with the highest number of publications co-authored with Yager. The table distinguishes between authors, organizations and countries.

Dimitar P. Filev is the most significant co-author of Yager with 33 joint publications. Five other authors have co-authored at least 10 publications with Yager and available in Web of Science. Regarding organizations, Iona College (USA) is the institution where Professor Yager has worked for most of his career. Therefore, most of his publications are under this affiliation. He also has a significant number of publications with King Saud University (Saudi Arabia) due to his recognition as

Table 4 Collaborators of Ronald R. Yager: Authors, organizations and countries

Authors	TP	Organizations	TP	Countries	TP
DP Filev	33	Iona College	527	USA	548
N Alajlan	19	King Saud U	23	Saudi Arabia	23
A Rybalov	16	U Alberta	11	Spain	17
FE Petry	12	US Navy	10	Italy	17
G Pasi	11	US Dep Defense	10	Canada	15
MZ Reformat	10	Naval Res Lab	10	France	12
V Kreinovich	8	National Science Found	9	Denmark	11
ZS Xu	6	CNRS France	8	PR China	10
HL Larsen	6	U Texas El Paso	7	UK	7
L Garmendia	6	Ford Motor Company	7	Israel	5
KJ Engemann	6	U California Berkeley	6	4 other	4
D Dubois	6	Pierre Marie Curie U Paris 6	6	Bulgaria	3
6 other	5	New York Academy Sci	6	3 other	2
4 other	4	8 other	5	8 other	1
15 other	3	6 other	4		
27 other	2	7 other	3		

Visiting Distinguished Scientist. The results in the country list are quite similar to organizations being the USA at the top of the list with most of his publications. Note that the publications from the seventies and eighties often do not include the affiliation. Due to this, the USA does not get all of his publications although this should be the result.

The research developed by Yager has influenced a lot of researchers. Tables 5 and 6 analyze the citing articles of Ronald Yager. Currently he has 9584 citing articles and 17523 citations. Thus, on average, each article cites Yager twice in the

Table 5 Citing articles of Ronald R. Yager: Authors, organizations and countries

Authors	TP	Organizations	TP	Countries	TP
RR Yager	415	Iona College	404	PR China	2234
ZS Xu	184	U Granada	306	USA	1408
JM Merigó	125	CNRS	194	Spain	1091
E Herrera-Viedma	111	Polish Academy of Sciences	178	France	546
J Kacprzyk	109	Southeast U	162	Taiwan	535
R Mesiar	102	Ghent U	129	UK	449
D Dubois	94	Slovak U Tech	115	Canada	400
F Herrera	92	U Toulouse	112	Poland	398
W Pedrycz	82	U Barcelona	109	Italy	355
GW Wei	78	U Toulouse III	102	India	314
L Martinez	78	U Jaén	102	Iran	307

(continued)

Table 5 (continued)

Authors	TP	Organizations	TP	Countries	TP
S Zadrozny	77	PLA U Sci Tech	95	Japan	254
H Bustince	76	Tsinghua U	93	Australia	240
H Prade	75	Islamic Azad U	91	Belgium	196
G Beliakov	61	Sichuan U	88	Turkey	164
J Montero	60	U Manchester	87	Brazil	162
V Torra	58	Indian Inst Tech	87	Germany	153
XW Liu	56	U Illes Balears	83	South Korea	150
J Torrens	55	Shanghai Jiao Tong U	83	Slovakia	139
F Chiclana	53	Public U Navarra	78	Czech Rep	124

reference list. Table 5 focuses on authors, organizations and countries citing the publications of Yager. Table 6 analyzes the journals and research areas where these citing articles are classified.

As expected, the self-citations of Yager are the most significant citing articles. This is quite logic because usually, his research has been built based on his previous studies. Moreover, Zeshui Xu, José M. Merigó, Enrique Herrera-Viedma, Janusz Kacprzyk and Radko Mesiar, have cited him in more than 100 publications each.

Table 6 Citing articles of Ronald R. Yager: Journals and research areas

Journals	TP	Research areas	TP
FSS	776	Computer science	7114
IS	457	Engineering	2683
IJIS	342	Mathematics	1488
IEEE-TFS	305	Operations Res. & Manag. Sci	930
Expert Syst with Applic	234	Automation & Control Syst	645
IJUFKS	229	Business & Economics	414
JIFS	212	Environmental Sci Ecology	180
IJAR	141	Imaging Sci Photograph Tech	146
KBS	139	Telecommunications	129
Eur J Operational Research	133	Sci Tech Other Topics	107
Applied Soft Computing	129	Robotics	105
IJGS	104	Water Resources	92
SC	93	Remote Sensing	90
IEEE-TSMC-B	74	Physics	83
IJ Computational Intel Syst	66	Mechanics	77
Computers & Industrial Eng	59	Instruments Instrumentation	70
Applied Math Modeling	54	Information Sci & Library Sci.	66
FODM	53	Materials Science	58
IEEE-TSMC-A	51	Transportation	54
IEEE-TSMC	50	Social Sci Other Topics	52

Organizations and countries are aligned with the affiliation of the authors of the citing articles. Among others, it is worth mentioning the University of Granada that has more than 300 citing articles. Regarding countries, China has the highest number of citing articles which proves the huge impact that Yager's research is having in this country. The USA and Spain also have more than 1000 citing articles.

Fuzzy Sets and Systems is the journal with the highest number of citing articles. Information Sciences also have a significant number of citations to Yager's research. Five other journals have more than 200 citing articles and five more have more than 100. Focusing on research areas, most of the citing articles are in the field of Computer Science. However, he also has a significant number of citations in Engineering, Mathematics and Operations Research & Management Science.

4   Bibliographic Coupling and Co-citations of Yager

In this Section, we visualize the publications of Yager and how they cite other research. For doing so, we use bibliographic coupling [7] and co-citation analysis [10]. Figure 3 presents the bibliographic coupling of authors which indicates the most significant co-authors of Yager and when do they tend to cite the same material.

Figure 4 presents the bibliographic coupling of countries based on Yager's publications. Note that the results are aligned with Fig. 3 because it represents the affiliation of these authors.

Figure 5 shows the bibliographic coupling of journals. Here, we can identify the journals where Yager publishes and see which ones tend to cite the same material.

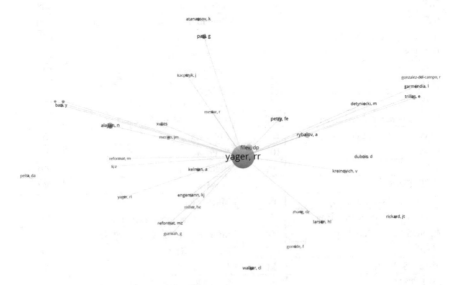

Fig. 3 Bibliographic coupling of authors

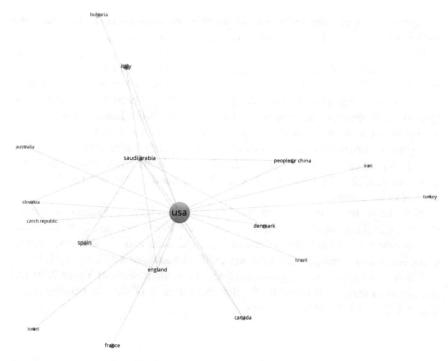

Fig. 4 Bibliographic coupling of countries

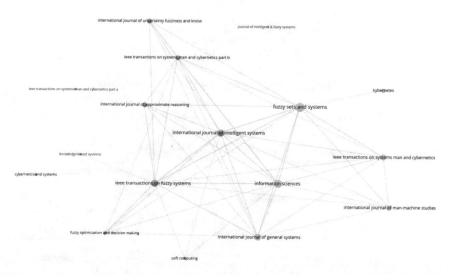

Fig. 5 Bibliographic coupling of journals

Figure 6 analyzes the bibliographic coupling of institutions. In this case, the results are also aligned with the affiliation of the authors of Fig. 3.

Next, let us focus on co-citation analysis. Figure 7 presents the author-co-citations of Yager. In summary, this figure shows the most cited authors by Yager and their connections when receiving citations by Yager's studies jointly. It is clear that he has cited himself mostly although he has been strongly influenced by Lotfi A. Zadeh, the father of fuzzy logic [15, 16]. Didier Dubois appears in the third position with also a significant influence in Yager's research.

Figure 8 analyzes the co-citations of documents that Yager cites in his publications. As expected, the most significant document is his seminal paper on OWA operators published in 1988. Most of the highly cited papers that he cite are by himself or by Zadeh.

Next, let us focus on the co-citations of journals. Figure 9 presents the results. Observe that the most cited journals by Yager are Fuzzy Sets and Systems, Information Sciences, IEEE Transactions on Systems, Man and Cybernetics, International Journal of Intelligent Systems and IEEE Transactions on Fuzzy Systems.

Finally, let us develop a keyword analysis of the publications of Yager. Note that this keyword analysis is based on the title and abstracts of the 585 publications of Yager. Figure 10 presents the results.

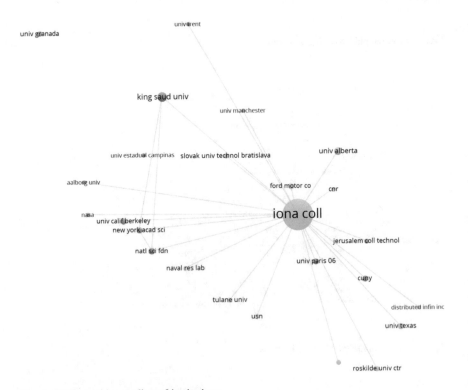

Fig. 6 Bibliographic coupling of institutions

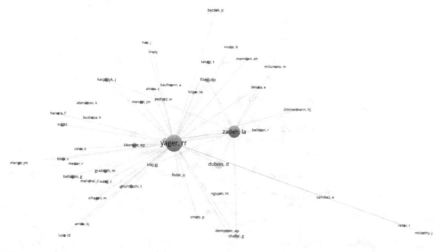

Fig. 7 Co-citations of authors

Fig. 8 Co-citations of documents

Apart from common words used in abstracts such as problems, approach, case and process, we can also identify some representative keywords of his research. First, we see at the right of the figure the keywords OWA, aggregation and operator, which clearly shows his contributions in this field. Some more general concepts such as information, knowledge and uncertainty are seen at the left side. At the top, appear several keywords connected with decision making. And at the bottom of the graph, we see the keywords connected to his contributions on fuzzy sets.

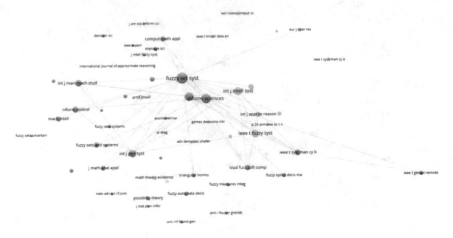

Fig. 9 Co-citations of journals

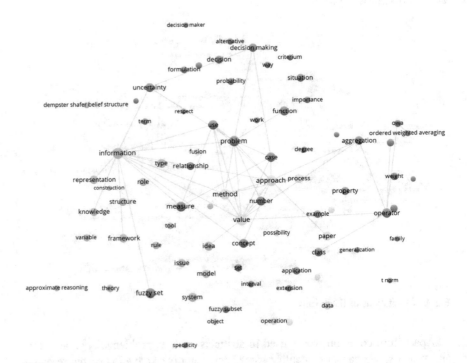

Fig. 10 Keyword analysis

5 Conclusions

This study has presented a general overview of the publications of Ronald R. Yager according to the results available in Web of Science. Yager is one of the World leading authors in Computer Science with more than 500 publications and 17000 citations. He is the editor-in-chief of the International Journal of Intelligent Systems and a member of the editorial board of many leading journals. This work has identified the journals and research areas where he usually publishes his research, his main collaborators, and the places where his research has influenced more.

Dimitar P. Filev is the most significant co-author throughout Yager's career. Yager tends to publish in Computer Research journals such as Fuzzy Sets and Systems, Information Sciences and some IEEE journals of the System, Man and Cybernetics Society and the Computational Intelligence Society. His most influential paper was written in 1988 and it is the first contribution on the OWA operators which has impacted significantly the scientific community. This study is among the 100 most cited papers of all-time in Computer Science. We have also developed a graphical analysis of the publications and citations of Yager through the VOS viewer software. By using bibliographic coupling and co-citation analysis we can map the leading variables that affect him. As expected, the graphical analysis provides similar results to those seen on the tables although we have seen some interesting differences. Particularly, we have seen the papers that Yager cites mostly which are mainly written by himself or by Lotfi A. Zadeh.

This study provides a general bibliometric overview of a leading author. This analysis provides a deep analysis of the main profile of a leading researcher which is quite useful to get an overview of a standard leading profile in this field. Obviously, each researcher has its own particularities but from a general perspective it is useful to get a general picture of the field. In future research, it would be interesting to develop similar bibliometric overviews with other authors in Computer Science and in other fields. This approach may provide a summary of a representative author that may be used by PhD students and newcomers in order to get a general overview of the field. It may be also useful in order to see in a paper a general summary of the contributions of an author. Although specialized researchers in this field could intuitively know this information, usually a bibliometric overview also identifies results that were not directly expected. Therefore, it may be useful to get a general perspective of an author.

As an interesting further direction of this analysis, one can use other systems, notably Scopus, which gain more and more relevance, and even the Goggle Scholar or systems based on it. This can be relevant for this analysis because the influence of Yager's works is very wide and concerns areas in which the WoS and journals covered in that system may not constitute the main part of publications covered. Moreover, these new systems take into account books which are very relevant elements of the publication record in many fields of science, exemplified by social sciences or humanities, which are not covered by the WoS but in which Yager's works have also been influential.

References

1. Alonso, S., Cabrerizo, F.J., Herrera-Viedma, E., Herrera, F.: H-index: a review focused on its variants, computation, and standardization for different scientific fields. J. Informetr. **3**, 273–289 (2009)
2. Bonilla, C., Merigó, J.M., Torres-Abad, C.: Economics in Latin America: a bibliometric analysis. Scientometrics **105**, 1239–1252 (2015)
3. Broadus, R.N.: Toward a definition of "Bibliometrics". Scientometrics **12**, 373–379 (1987)
4. Cobo, M.J., Martínez, M.A., Gutiérrez-Salcedo, M., Fujita, H., Herrera-Viedma, E.: 25 years at knowledge-based systems: a bibliometric analysis. Knowl.-Based Syst. **80**, 3–13 (2015)
5. Emrouznejad, A., Marra, M.: Ordered weighted averaging operators 1988–2014. a citation based literature survey. Int. J. Intell. Syst. **29**, 994–1014 (2014)
6. Hirsch, J.E.: An index to quantify an individual's scientific research output. Proc. Natl. Acad. Sci. U.S.A. **102**, 16569–16572 (2005)
7. Martyn, J.: Bibliographic coupling. J. Doc. **20**, 236 (1964)
8. Merigó, J.M., Gil-Lafuente, A.M., Yager, R.R.: An overview of fuzzy research with bibliometric indicators. Appl. Soft Comput. **27**, 420–433 (2015)
9. Merigó, J.M., Mas-Tur, A., Roig-Tierno, N., Ribeiro-Soriano, D.: A bibliometric overview of the Journal of Business Research between 1973 and 2014. J. Bus. Res. **68**, 2645–2653 (2015)
10. Small, H.: Co-citation in the scientific literature: a new measure of the relationship between two documents. J. Am. Soc. Inform. Sci. **24**, 265–269 (1973)
11. Van Eck, N.J., Waltman, L.: Software survey: VOSviewer, a computer program for bibliometric mapping. Scientometrics **84**, 523–538 (2010)
12. Yager, R.R.: On ordered weighted averaging aggregation operators in multi-criteria decision making. IEEE Trans. Syst. Man Cybern. **18**, 183–190 (1988)
13. Yager, R.R.: Families of OWA operators. Fuzzy Sets Syst. **59**, 125–148 (1993)
14. Yager, R.R., Kacprzyk, J., Beliakov, G.: Recent Developments on the Ordered Weighted Averaging Operators: Theory and Practice. Springer-Verlag, Berlin (2011)
15. Zadeh, L.A.: Fuzzy sets. Inf. Control **8**, 338–353 (1965)
16. Zadeh, L.A.: Is there a need for fuzzy logic? Inf. Sci. **178**, 2751–2779 (2008)